Linear
Rational Expectations
Models
A User's Guide

Linear Rational Expectations Models

A User's Guide

Charles H. Whiteman
Department of Economics
The University of Iowa
Iowa City, Iowa

University of Minnesota Press □ Minneapolis

Copyright © 1983 by the University of Minnesota

All rights reserved.

Published by the University of Minnesota Press,

2037 University Avenue Southeast, Minneapolis MN 55414

Printed in the United States of America

Library of Congress Cataloging in Publication Data

Whiteman, Charles H.
   Linear rational expectations models.

   Bibliography: p.
   Includes indexes.
   1. Economics—Mathematical models. 2. Rational
expectations (Economic theory) I. Title.
HB141.W55 1983   330'.0724   83-1280
ISBN 0-8166-1181-5
ISBN 0-8166-1179-3 (pbk.)

The University of Minnesota

is an equal-opportunity

educator and employer.

To my parents

Preface

The rational expectations critique of conventional policy analysis centers on the failure of traditional models to embody the notion that the behavior of consumers and producers changes when the economic environment in which they operate changes. The rational expectations challenge is to build models of the economy in which agents' views about the future are explicitly tied to the objective features of the environment in which they operate. The challenge is, to date, largely unmet; few such models have yet been constructed. The reason for this is that the assumption that agents respond rationally to changes in the economic environment introduces complicated new restrictions among the equations of the economic model. These restrictions, "the hallmark of rational expectations models," result in a number of technical problems which must be solved before the computational burden imposed by rational expectations can be reduced to a manageable level.

The analysis herein concerns the solution of stationary, linear rational expectations models, where the "solution" to such a model is actually a particular sort of fixed point. While methods exist for calculating these fixed points for a certain class of linear models, the techniques presented here are, I think, simpler and more broadly applicable.

Techniques for finding solutions in rational expectations models are, in effect, sophisticated guessing schemes. One makes a shrewd guess at the form of the solution, and then uses the guess to determine the values of the parameters of the solution. The "undetermined coefficient" technique used here involves a transformation of the model from the "time-domain" to the "frequency-domain." In so doing, it allows the model to be viewed from a different and potentially very useful perspective.

Before one begins the search for a solution to a rational expectations model, it is useful to specify where one will look. The principle which guides the technique presented below is that solutions--expectations equilibria--should be functions of the information possessed by the agents of the model. This principle and the frequency-domain solution technique are presented and applied in a number of settings in Chapter I. A comparison of this technique with other existing methods in Chapter II illustrates the advantages of the frequency-domain technique. A further illustration is provided in Chapter III, which studies the possible nonuniqueness of rational expectations equilibria.

In Chapters IV and V, the frequency-domain technique is applied in two contexts which have proved difficult to analyze using conventional methods. First, (nearly) closed-form expressions for solutions to multivariate rational expectations models are obtained in Chapter IV. This opens up the possibility that large-scale linear econometric models can be estimated and manipulated under the rational expectations hypothesis. Second, the technique is used in Chapter V to establish conditions for the existence and uniqueness of equilibria in a class of dynamic models. Given existence, a simple application of the technique allows closed-form computation of equilibria heretofore calculable only by iterative methods.

Elementary results of the residue calculus and the theory of stationary stochastic processes are exploited throughout this work. The book by Churchill, Brown, and Verhey (1974) is a good reference on the complex analysis to be used; on stochastic processes, see Whittle (1963).

Chapters I-IV are adapted from my Ph.D. dissertation, "Moving Average Representations in Rational Expectations Models," submitted to the University of Minnesota. I have received helpful comments as well as various versions of the dissertation (folk-) theorem, "nothing is impossible to the person who doesn't have to do it himself," from Ian Bain, Christopher Bingham, Martin Eichenbaum, John Geweke, Lars Hansen, Takatoshi Ito, John Kennan, Bennet McCallum, Michael Salemi, Christopher Sims, and Neil Wallace. But my greatest debt is to my advisor, Thomas Sargent. I first learned about rational expectations from him, and he either influenced or inspired much of what appears in the following pages.

I would like to thank the Instructional Support Service Office of the College of Business at The University of Iowa for helping me produce this book. Special thanks go to Cindy Allbaugh, who typed the entire manuscript. Finally, thanks go to Lindsay Waters, Beverly Kaemmer, and the staff of The University of Minnesota Press for valuable assistance.

Charles H. Whiteman
Iowa City, Iowa

Contents

Preface   vii

CHAPTER I   MOVING AVERAGE REPRESENTATIONS FOR SCALAR RATIONAL
             EXPECTATIONS MODELS   1

             1. The First Order Case   4
             2. The Second Order Case   12
             3. The nth Order Case   16
             4. Bivariate Representations   22
             5. Perturbed Equations   26
             6. Withholding Equations   29
                Conclusion   36

CHAPTER II   SOLUTION TECHNIQUES FOR RATIONAL EXPECTATIONS MODELS:
             A CRITICAL REVIEW   37

             1. A Benchmark Solution Technique   39
             2. Time Domain: Moving Averages   44
             3. Time Domain: Autoregressions   50
             4. Forward and Backward Solutions   55
             5. Operator Techniques   61
             6. The Saracoglu-Sargent Technique   63
                Conclusion   69

CHAPTER III CONFRONTING NONUNIQUENESS IN RATIONAL EXPECTATIONS
             MODELS   71

             1. A Simple Expectational Difference Equation:
                Parametric Nonuniqueness   72
             2. The "Leading Indicator" Problem   74
             3. The Minimum Variance Condition   78
             4. The Minimum State Variable Technique   84

	5. A Restrictive Solution Concept	86
	Conclusion	88
CHAPTER IV	MOVING AVERAGE REPRESENTATIONS FOR MULTIVARIATE RATIONAL EXPECTATIONS MODELS	90
	1. An Existence and Uniqueness Theorem	90
	2. An Extension	100
	Conclusion	101
	Appendix	103
CHAPTER V	A SET OF OPTIMIZATION EXAMPLES	110
	1. The Model	111
	2. Perfectly Elastic Supply and Demand	114
	3. Inelastic Demand; Perfectly Elastic Supply	116
	4. Feedback from Decisions to States	119
	Conclusion	123

References 125

Author Index 130

Subject Index 132

Linear
Rational Expectations
Models
A User's Guide

I

Moving Average Representations for Scalar Rational Expectations Models

Although superficially similar, the equations to be studied in this book are quite unlike ordinary stochastic difference equations: each contains a term involving an expectation. For this reason, the objects to be studied are referred to as "expectational difference equations." This nomenclature distinguishes the objects of study from ordinary stochastic difference equations in a way that Shiller's (1978) term "linear difference models" does not.

The simplest example of an expectational difference equation is

(1) $\quad E_t y_{t+1} - \rho y_t = x_t. \qquad \rho \in R, \rho \neq 0$

For most purposes, t indexes the integers, and x_t and y_t may be thought of as representative elements in sequences of random variables. The variable x_t is referred to as "exogenous," or as "the driving process," while the expression "$E_t y_{t+1}$" means "the time t forecast of y_{t+1}."

If forecasts are taken to be arbitrary fixed-weight mark-ups of past values--if expectations are "adaptive"--then (1) is really just an ordinary difference equation: substituting $\sum_{j=0}^{m} \alpha_j y_{t-j}$ for $E_t y_{t+1}$,

1

(2) $\quad \sum_{j=0}^{m} \alpha_j y_{t-j} - \rho y_t = x_t.$

Methods for finding "solutions"--expressions giving y_t solely in terms of $\{x_t, t \in I\}$--to such equations are well known. In particular, suppose that $\alpha_1 = 1$, $\alpha_j = 0$, $j \neq 1$. Then (2) becomes

(3) $\quad y_{t-1} - \rho y_t = x_t.$

Without any side conditions, there are a number of solutions to (3). For instance, by repeatedly using (3) to calculate y_{t-j} for $j > 0$ and substituting the result back into (3), one obtains the "backward solution"

$$y_t^1 = \rho^{-1} \sum_{j=0}^{\infty} \rho^{-j} x_{t-j}.$$

Since $\rho^{-(t-1)} - \rho(\rho)^{-t} = 0$, $y_t^2 = y_t^1 + \tilde{c}\rho^{-t}$ is also a solution for any real number $\tilde{c}$. On the other hand, by repeatedly using (3) to calculate y_{t+j} for $j > 0$ and substituting the result back into (3), one obtains the "forward solution"

$$y_t^3 = \sum_{j=0}^{\infty} \rho^j x_{t+j+1}.$$

Again, $y_t^4 = y_t^3 + \tilde{d}\rho^{-t}$ is also a solution for any $\tilde{d} \in R$.

It is important to note that y_t^i $i = 1,\ldots,4$ were obtained <u>formally</u> by manipulating (3). Clearly, there is no guarantee that either of the infinite series given by y_t^1 and y_t^3 converges. Following Sargent (1979b, Chapter IX), suppose that $\{x_t: t \in I\}$ is a bounded sequence. Then $\{y_t^1: t \in I\}$ is a bounded sequence when $|\rho| > 1$ and $\{y_t^3: t \in I\}$ is a bounded sequence when $|\rho| < 1$. If $\{y_t: t \in I\}$ is to be bounded for all t, one chooses either y_t^1 or y_t^3; y_t^2 and y_t^4 diverge as $t \to \infty$ ($-\infty$) for $|\rho| < 1$ (>1).

The preceding analysis suggests that given a structure on $\{x_t\}$ (boundedness), it is useful to study solutions with the same general structure. That is, it is useful to search for solutions in the same "space"--that of bounded sequences--as the driving process. The reason is that without any restrictions on $\{x_t\}$ and without any side conditions, any linear combination of y_t^2 and y_t^4 formally "solves" (3). Yet, many economic models of the form (1) have been proposed without any such restrictions or side conditions.

The central message of Lucas's (1976) critique of econometric policy evaluation is that the structure of driving processes is of vital importance. In addition, the many results obtained by Hansen and Sargent (1980a, 1980b, 1980c, 1981a), in situations where they are _forced_ to examine solutions of the same general nature as driving processes, suggest that _choosing_ to so restrict one's attention in other contexts may also prove useful.

The solution strategy implicit in the preceding paragraph will be followed throughout this book. It uses a powerful mathematical technique called the "z-transform"; the following statement makes it concrete.

Solution Principle:

1. Driving processes are taken to be zero-mean linearly regular covariance stationary stochastic processes with known Wold representations.

2. Expectations are formed rationally, and are computed using the Wiener-Kolmogorov formulas.

3. Solutions will be sought in the space spanned by time-independent square-summable linear combinations of the process fundamental for the driving process.[1]

4. The rational expectations restrictions will be required to hold for all realizations of the driving process.

---

[1]When $\{\varepsilon_t : t \in J\}$ is fundamental for the driving process, such solutions are of the form $\{y_t^I\} = \{\sum_{j=0}^{\infty} c_j \varepsilon_{t-j}\}$, where $\sum_{j=0}^{\infty} c_j^2 < \infty$. Time-dependent

It is this Solution Principle which distinguishes this work from other studies of the solutions to expectational difference equations. Such studies, including Blanchard (1978), Shiller (1978), Aoki and Canzoneri (1979), Blanchard and Kahn (1980), McCallum (1980), and Gourieroux, Laffont, and Monfort (1982), depart in some way from one or more of the four tenets of the Solution Principle.

The analysis in the remainder of this chapter applies the Solution Principle to the study of scalar expectational difference equations. First, second, and n^{th} order equations are discussed in sections 1, 2, and 3. A strategy for computing restricted bivariate representations is presented in section 4. Section 5 deals with solutions to equations like (1) but which are perturbed by stochastic processes with well-defined structures. "Withholding" equations--those with information sets and "driven" variables dated differently--are discussed in section 6.

1. The First Order Case

Examples of first order expectational difference equations like (1) appear frequently in the macroeconomics literature. By far the most studied (Sargent and Wallace (1973), Jacobs (1975), Sargent (1976), Sargent (1977), Blanchard (1978), Saracoglu and Sargent (1978), Salemi and Sargent (1979), Salemi (1979), Christiano (1980), McCallum (1980), and Gourieroux, Laffont,

---

solutions are of the form $\{y_t^D\} = \{\sum_{j=0}^{\infty} C_{j,t} \varepsilon_{t-j}\}$. The coefficients in the expression for y_t^D depend on the lag (j) as well as the absolute position in time (t), while those in the expression for y_t^I depend only on the lag. When $C_{j,t}$ depends nontrivially on t, $\{y_t^D\}$ is not stationary, though to distinguish this from another type of nonstationarity (explosiveness), $\{y_t^D\}$ is sometimes said to be "unstable."

and Monfort (1982)) equation is Cagan's (1956) portfolio balance equation

$$m_t - p_t = \alpha(E(p_{t+1}|I_t) - p_t),$$

where p_t is the log of the price level at time t, I_t is an information set available at t, m_t is the log of the money stock at t, and α is a parameter. By analogy to (1), m_t is the driving process and p_t is the driven variable. The expression $E(p_{t+1}|I_t)$ is generally taken to mean the best forecast of p_{t+1} based linearly on the elements of I_t, a notion to be used throughout this book. Note that the information set I_t has the same date as the most recent (known) value of the driven variable, p_t. This makes Cagan's equation an expectational difference equation rather than a withholding equation (section 6 below.) The sign and magnitude of α determine the form of the solution for p_t in terms of $\{m_s, s \leq t\}$. The simplest way to see this is to abstract from the application and return to (1):

(1) $\qquad E_t y_{t+1} - \rho y_t = x_t.$ $\qquad\qquad\qquad \rho \in R, \rho \neq 0$

According to the first tenet of the Solution Principle discussed above, some structure must be placed on $\{x_t\}$. For the purposes of this analysis, x_t will be taken to be a zero-mean linearly regular covariance stationary stochastic process (LRCSSP^0). Thus x_t has the Wold representation

(4) $\qquad x_t = \sum_{j=0}^{\infty} A_j \varepsilon_{t-j},$

with $\varepsilon_t = x_t - E(x_t|x_{t-1}, x_{t-2}, \ldots)$, $\Sigma A_j^2 < \infty$, and the function $A(z) = \sum_{j=0}^{\infty} A_j z^j$ must be analytic on the open unit disk. Thus (4) will often be written as

(4') $\qquad x_t = A(L)\varepsilon_t$

where L is the lag operator: $L^n x_t = x_{t-n}$. Because $\{\varepsilon_t\}$ is fundamental for $\{x_t\}$, the (Hilbert) space generated (spanned) by square-summable linear combinations of $\{\varepsilon_s, s \leq t\}$ is the same as that spanned by $\{x_s, s \leq t\}$. Following Rozanov (1967), the completions in mean square norm of these spaces will be referred to as $H_{\varepsilon}(t)$ and $H_x(t)$.

By tenet three of the Solution Principle, solutions--$\{y_t\}$ processes--will be sought in $H_x(t)$. Thus a solution, y_t, satisfies (1) and is representable by a square-summable linear combination of $\{\varepsilon_s, s \leq t\}$. The sequence $\{\varepsilon_t\}$ is fundamental for such solutions, which can therefore be written as

$$(5) \quad y_t = \sum_{j=0}^{\infty} C_j \varepsilon_{t-j} = C(L)\varepsilon_t$$

with $\Sigma C_j^2 < \infty$. Expression (5) is the non-normalized Wold representation for y_t. Thus $C(z)$ must be analytic on the open unit disk.

The information set implicit in the forecast $E_t y_{t+1}$ is given by $\{x_t, x_{t-1}, \ldots\}$ or, from (4), $\{\varepsilon_t, \varepsilon_{t-1}, \ldots\}$. Thus, using (5), the Wiener-Kolmogorov formulas give

$$E_t y_{t+1} = E_t [C_0 \varepsilon_{t+1} + C_1 \varepsilon_t + C_2 \varepsilon_{t-1} + \ldots]$$

$$= C_1 \varepsilon_t + C_2 \varepsilon_{t-1} + \ldots$$

$$= \sum_{j=0}^{\infty} C_{j+1} \varepsilon_{t-j}.$$

But (1) requires that

$$(6) \quad \sum_{j=0}^{\infty} C_{j+1} \varepsilon_{t-j} - \rho \sum_{j=0}^{\infty} C_j \varepsilon_{t-j} = \sum_{j=0}^{\infty} A_j \varepsilon_{t-j},$$

which, by tenet 4, must hold for all realizations of $\{\varepsilon_t\}$. This requires that the sequences of Wold moving average coefficients represented in (6) be equal, or, equivalently, that their z-transforms be identical as analytic functions on the open unit disk. The z-transform of the sequence in the first term of (6) is given by (the Wiener-Kolmogorov formula) $[C(z)/z]_+ = z^{-1}(C(z) - C_0)$. The transform of $\{C_0, C_1, \ldots\}$ is $C(z)$, and that of $\{A_0, A_1, \ldots\}$ is $A(z)$. Thus, (1) requires that $z^{-1}(C(z) - C_0) - \rho C(z) = A(z)$. Multiplying by z and rearranging gives

$$(1-\rho z)C(z) = zA(z) + C_0$$

or

(7) $\qquad C(z) = (1-\rho z)^{-1}\{zA(z) + C_0\}$

which, formally, is the z-transform of the sequence of coefficients $\{C_0, C_1, \ldots\}$ in the Wold representation for y_t. But (7) is analytic for $|z| < 1$ if and only if $|\rho| \leq 1$; otherwise, $C(z)$ has a removable singularity at $\rho^{-1}\varepsilon\{z: |z| < 1\}$.

When $|\rho| > 1$, the free parameter C_0 can be set in such a way to make $C(z)$ analytic at ρ^{-1} using an extension of the methods introduced by Futia (1981). What is required is that the residue of $C(\cdot)$ at ρ^{-1} be zero:

$$\lim_{z \to \rho^{-1}} (1-\rho z)C(z) = \{\rho^{-1}A(\rho^{-1}) + C_0\} = 0,$$

giving $C_0 = -\rho^{-1}A(\rho^{-1})$. The parameter C_0 can be defined this way provided $A(\rho^{-1})$ is well-defined, a condition guaranteed by the square-summability assumption made about the moving average coefficients of the process $\{x_t\}$. Thus, for $|\rho| > 1$,

(8) $\qquad C(z) = (1-\rho z)^{-1}\{zA(z) - \rho^{-1}A(\rho^{-1})\}.$

From (7) and (8), the Wold representation for y_t, restricted by (1), can be written

(9) $\quad y_t = (1-\rho L)^{-1}\{LA(L) + C_0\}\varepsilon_t \qquad |\rho| < 1,$

$\qquad\quad = (1-\rho L)^{-1} x_t + C_0(1-\rho L)^{-1}\varepsilon_t$

or

(10) $\quad y_t = (1-\rho L)^{-1}\{LA(L) - \rho^{-1}A(\rho^{-1})\}\varepsilon_t \qquad |\rho| > 1.$

If $A(L)$ has an inverse in nonnegative powers of L, a condition equivalent to the requirement that x have an autoregressive representation, (9) and (10) can be used to express y_t in terms of current and past values of x_t:

(9') $\quad y_t = (1-\rho L)^{-1}\{L + C_0 A(L)^{-1}\} x_t \qquad |\rho| < 1,$

(10') $\quad y_t = (1-\rho L)^{-1}\{L - \rho^{-1}A(\rho^{-1})A(L)^{-1}\} x_t \qquad |\rho| > 1.$

The explicit dependence of the parameters of (10') on the parameters of the Wold representation of $\{x_t\}$ is what Sargent (1981) has referred to as "the hallmark of rational expectations": cross-equation restrictions. The coefficients on $x_t, x_{t-1}, \ldots$ in these expressions can be found by using the polynomial long-division algorithm described by Sargent (1979b).

There are a number of points to be made about the solutions represented by (9) and (10). First, and somewhat trivially, both satisfy the four tenets of the Solution Principle.

Second, for $|\rho| < 1$, there are many solutions $\{y_t\}$ with representative element y_t in $H_x^-(t)$. The reason for this is that C_0 is a free parameter in (9): for $|\rho| < 1$, (9) gives a family of sequences $\{y_t\}$ indexed by C_0, each of whose members solves (1) and lies in the space spanned by current and past values of $\{x_t\}$. Any finite value of C_0 is as good as any other: (1) does not uniquely determine the Wold representation of $\{y_t\}$. This nonuniqueness of solutions has been encountered by numerous authors, including Black (1974), Brock (1975), and Taylor (1977). Although the initial nonuniqueness results arose in models of money under rational expectations or perfect foresight, it should be clear that the nonuniqueness of (9) has nothing to do with money. The nonuniqueness is a mathematical consequence of (1) together with $|\rho| < 1$.

Third, there is no natural way to resolve the nonuniqueness. For example, there is <u>no</u> reason to set $C_0 = -\rho^{-1}A(\rho^{-1})$ when $|\rho| < 1$, as Gourieroux, Laffont, and Monfort (1982) do. Simply put, the nonuniqueness of (9) cannot be resolved without imposing some side conditions. A number of these side conditions are studied in Chapter III.

Fourth, when $|\rho| > 1$, (10) gives the unique solution y_t lying in $H_x^-(t)$; that is, the stochastic process $\{y_t\}$ which is informationally equivalent to $\{x_t\}$. In this case, the structure of $\{x_t\}$, (4), and the restriction (1) uniquely determine the pattern in the Wold moving average coefficients C_0, $C_1, \ldots$ It is the relation between C_i and C_j, not the absolute size of these coefficients, that is determined by (4), (1), and the Solution Principle: $y_t = C(L)\varepsilon_t$ with $C_0 = -\rho^{-1}A(\rho^{-1})$ can be renormalized as follows. Write $y_t = D(L)\varepsilon_t^*$ where $D_j = C_0^{-1}C_j$ and $\varepsilon_j^* = C_0\varepsilon_t$. Clearly, the same rescaling procedure can be applied to the representation for x_t. Of course, any other constant divisor would work as well. This normalization procedure is related to one of the identification problems discussed by Hansen and Sargent (1981b).

Fifth, it is important to note that there are solutions y_t which do not lie in $H_x^-(t)$. For instance, when $|\rho| < 1$,

(11) $\qquad y_t = (1-\rho L)^{-1}\{LA(L) + C_0\}\varepsilon_t + \tilde{c}\rho^t$

for any $\tilde{c} \in R$ solves (1). But ρ^t is a deterministic function of t; the solution in (11), unlike $\{x_t\}$, is not indeterministic.

Finally, the operator in (10), $(1-\rho L)^{-1}\{LA(L) - \rho^{-1}A(\rho^{-1})\}$ must be considered as a whole. Although it represents a function which is analytic on the open unit disk, one of its components, $(1-\rho L)^{-1}$, does not. Indeed, suppose $|\rho| > 1$. Then

(12) $\quad y_t = (1-\rho L)^{-1}\{LA(L) + C_0\}\varepsilon_t \quad C_0 \neq -\rho^{-1}A(\rho^{-1})$

yields a solution $\{y_t\}$ which cannot be expressed by square-summable linear combinations of current and past values of $\{x_t\}$. The reason is that the sequence $C_0, C_1, \ldots$ represented by $(1-\rho L)^{-1}\{LA(L) + C_0\}$ is not square-summable as a result of the isolated singularity at ρ^{-1} in its z-transform. Clearly, there is a family of such solutions indexed by C_0. Thus for $C_0 \neq -\rho^{-1}A(\rho^{-1})$, the solution (12) does not lie in $H_{\bar{x}}(t)$. But does this mean that $\{y_t\}$ is nonstationary? Only, to almost quote Sims (1977, p. 31), if one imposes the additional requirement that (12) be causal with x as input and y as output. Put differently, (12) implies a nonstationary y only if it is required that $\{y_s: s \leq t\}$ not span a "larger" space than $\{\varepsilon_s: s \leq t\}$.

The reason for this is that under a natural interpretation, (12) is not evidence of nonstationarity, it is just not a fundamental representation. To see this, write $w_t = \{C_0 + LA(L)\}\varepsilon_t = C_0\varepsilon_t + x_{t-1}$ and note that (12) is (formally) implied by

(13) $\quad w_t = (1-\rho L)y_t$.

For finite C_0, w_t can be taken to be a LRCSSP^0. But y_t does not lie in $H_{\bar{w}}(t)$. It need not, however, be nonstationary. For example, take $\{x_t\}$ to be white noise. Then (13) is a representation, though not a fundamental one, for the stationary $\{w_t\}$. The fundamental representation for w_t is

$w_t = \theta(L)y_t^*$ where $\theta(z)\theta(z^{-1}) = (1-\rho z)(C_0+z)(C_0+z^{-1})(1-\rho z^{-1})$ for $|z| = 1$ and $y_t^* = w_t - E(w_t | w_{t-1}, w_{t-2}, \ldots)$. The moving average coefficients are represented by $\theta(z) = (1-\rho z)[(1-\rho_1^{-1}z)/(z-\rho^{-1})](C_0+z)$, where the expression in brackets is one of Blaschke's factors.[2] (See Hansen and Sargent (1980a).) It should be clear that none of this analysis requires that $\{y_t\}$ be nonstationary, though it does require that the system (y_t, x_t) be "anti-causal": y causes x.

The preceding analysis suggests why the search for solutions lying in the space spanned by driving processes is useful and important. Simply put, if there is one solution lying in another space, there is an infinity of such solutions.

The multiplicity of solutions in other spaces is not peculiar to stochastic difference equations; it is not even peculiar to expectational difference equations with rational expectations. It is a property of difference equations in general. The problem should cause economists no more discomfort than it does an engineer. But McCallum (1980, note 1) documents no small amount of consternation:

> Shiller, for example, says "The existence of so many solutions to the rational expectations model implies a fundamental indeterminacy for these models" (1978, p. 33). Blanchard states that "in models where anticipations of future endogenous variables influence current behavior, there exists an infinity of solutions under the assumption of rational expectations" (1979, p. 114). Burmeister [1980] suggests that "one of the most crucial issues in rational expectations modelling ... concerns the dynamic properties of rational expectations paths and the manner in which the stability properties of these expectations serves to make determinate the stochastic properties of the actual variables" [1980, p. 800].

The reason these authors are concerned is that each, in one way or another, suggests as solutions to (1) linear combinations of objects like (9), (11),

---

[2] $|C_0| > 1$ was assumed for expositional convenience. When $|C_0| < 1$, $\theta(z)$ comprises two of Blaschke's factors.

and (12), objects in different spaces. These procedures are discussed in Chapter III.

Linear rational expectations models are generally constructed to address data. It seems natural, therefore, to restrict one's attention, at least for the present, to solutions susceptible to analysis by standard econometric techniques. For without some restrictions, economists will never get anywhere with *any* models, let alone rational expectations models. The natural solutions, therefore, are the ones lying in the space spanned by driving processes. It will be seen in the next section, however, that there need not be *any* such solutions.

2. The Second Order Case

Second order expectational difference equations sometimes arise as primitive concepts in models postulated at the level of demand functions. For instance, they arise in discrete time versions of the Dornbusch (1976) and Wilson (1979) studies of exchange rate dynamics. More often, perhaps, such equations arise as necessary conditions for optima in linear-quadratic versions of costly adjustment models. These "Euler equations," studied in Chapter V, can be found in Sargent (1978, 1979b), Kennan (1979), Hansen and Sargent (1980a, 1980b, 1980c, 1981b), Eckstein (1980), and Eichenbaum (1981). The simplest example of such an equation is

(14) $E_t y_{t+1} - (\rho_1 + \rho_2) y_t + \rho_1 \rho_2 y_{t-1} = x_t$.

As in section 1, the Solution Principle will be applied to (14). First, assume that $\{x_t\}$ has the Wold representation

(15) $x_t = A(L)\varepsilon_t$

and that $\{y_t\}$ can be represented as

(16) $\quad y_t = C(L)\varepsilon_t.$

Then (14) requires that

(17) $\quad \sum_{j=0}^{\infty} C_{j+1}\varepsilon_{t-j} - (\rho_1+\rho_2) \sum_{j=0}^{\infty} C_j\varepsilon_{t-j} + \rho_1\rho_2 \sum_{j=0}^{\infty} C_j\varepsilon_{t-j-1}$

$= \sum_{j=0}^{\infty} A_j\varepsilon_{t-j}.$

As in section 1, the z-transforms of the sequences represented in (17) must be identical as analytic functions on the open unit disk:

$$z^{-1}(C(z) - C_0) - (\rho_1 + \rho_2)C(z) + \rho_1\rho_2\, zC(z) = A(z)$$

or, multiplying by z and rearranging,

$$(1-\rho_1 z)(1-\rho_2 z)C(z) = zA(z) + C_0.$$

Formally, then,

(18) $\quad C(z) = \{(1-\rho_1 z)(1-\rho_2 z)\}^{-1}\{zA(z) + C_0\}.$

There are three cases to be considered.

First, suppose that $|\rho_1| < 1$, $|\rho_2| < 1$. Then (18) describes a function $C(z)$ which is analytic on $|z| < 1$. The Wold representation for $\{y_t\}$ is then

(19) $\quad y_t = \{(1-\rho_1 L)(1-\rho_2 L)\}^{-1}\{LA(L) + C_0\}\varepsilon_t$

which can be used to write y_t in terms of $\{x_s: s \leq t\}$ as

$$y_t = \{(1-\rho_1 L)(1-\rho_2 L)\}^{-1}\{L+C_0 A(L)^{-1}\}x_t,$$

provided that $\{x_t\}$ has an autoregressive representation. In this case, the restriction (14) does not uniquely determine the solution $\{y_t\}$. For any finite value of C_0, (19) gives a process $\{y_t\}$, lying in $H_{\bar{x}}(t)$, which satisfies (14). This situation is identical to the one encountered in section 1 when $|\rho| < 1$. But notice that (19), like (9), <u>is</u> restrictive. Though the sequence of moving average coefficients represented in (19) generally has many non-zero elements, only one, C_0, is left undetermined by (14) and (15). Of course, <u>each</u> of the moving average coefficients for $\{y_t\}$ depends on this free parameter; the point is that the z-transform of this sequence can be written with only one unknown.

The second case is $|\rho_1| < 1 < |\rho_2|$. With this parameterization, the function $C(z)$ in (18) has an isolated singularity at ρ_2^{-1}. But $C(\cdot)$ can be made analytic on $|z| < 1$ by setting C_0 in such a way as to cause the residue of $C(\cdot)$ at ρ^{-1} to equal zero:

$$\lim_{z \to \rho_2^{-1}} (1-\rho_2 z)C(z) = (1-\rho_1\rho_2^{-1})^{-1}\{\rho_2^{-1}A(\rho_2^{-1}) + C_0\} = 0.$$

Thus $C_0 = -\rho_2^{-1}A(\rho_2^{-1})$, an expression similar to that obtained in section 1 with $|\rho| > 1$. Thus, with $C_0 = -\rho_2^{-1}A(\rho_2^{-1})$, (18) can be used to write the Wold representation for $\{y_t\}$ as

(20) $\quad y_t = \{(1-\rho_1 L)(1-\rho_2 L)\}^{-1}\{LA(L) - \rho_2^{-1}A(\rho_2^{-1})\}\varepsilon_t.$

As above, (20) can be used to write y_t in terms of current and past values of x_t:

$$y_t = \{(1-\rho_1 L)(1-\rho_2 L)\}^{-1}\{L-\rho_2^{-1}A(\rho_2^{-1})A(L)^{-1}\}x_t.$$

In this case, except for normalization, (14) and (15) uniquely determine a fundamental representation for y_t. As in section 1, when $C_0 \ne -\rho_2^{-1}A(\rho_2^{-1})$, there is a stationary y_t, not in $H_{\bar{x}}(t)$, whose fundamental representation is not restricted by (14).

The third case, $1 < |\rho_1|$, $1 < |\rho_2|$, illustrates the importance of studying the second order expectational difference equation. In this situation $C(z)$ has two isolated singularities, at ρ_1^{-1} and ρ_2^{-1}. Suppose that $\rho_1 \neq \rho_2$. Then for $C(z)$ to be analytic on $|z| < 1$, <u>both</u>

$$\lim_{z \to \rho_1^{-1}} (1-\rho_1 z)C(z) = (1-\rho_2\rho_1^{-1})^{-1}\{\rho_1^{-1}A(\rho_1^{-1}) + C_0\} = 0$$

and

$$\lim_{z \to \rho_2^{-1}} (1-\rho_2 z)C(z) = (1-\rho_1\rho_2^{-1})^{-1}\{\rho_2^{-1}A(\rho_2^{-1}) + C_0\} = 0.$$

But this requires that $C_0 = -\rho_1^{-1}A(\rho_1^{-1})$ and $C_0 = -\rho_2^{-1}A(\rho_2^{-1})$, which is impossible when $\rho_1 \neq \rho_2$. The problem does not disappear when $\rho_1 = \rho_2$. In this case, the residue of $C(z)$ at ρ_1 cannot be calculated using the above formulas. Rather, the residue of $C(\cdot)$ at ρ_1 is

$$\frac{d}{dz}A(z)\Big|_{z = \rho_1^{-1}} + A(\rho_1^{-1})$$

which is not generally zero. In particular, C_0 cannot be set to make $C(\cdot)$ analytic on $|z| < 1$. Hence, when $1 < |\rho_1|$ and $1 < |\rho_2|$ there is no solution $\{y_t\}$ to (14) with elements y_t lying in $H_x^-(t)$. Equation (14) is overly restrictive in the sense that it drives the researcher out of $H_x^-(t)$ and into some other space where, as has been seen above, (14) is not restrictive.

First order equations like (1) with complex coefficients generally do not arise. But if

$$E_t y_{t+1} - \phi y_t + \psi y_{t-1} = x_t,$$

the factorization written in (14) may well yield complex ρ_1 and ρ_2, one being the conjugate of the other. In such situations, $|\rho_1| = |\rho_2|$ and either the first or third cases discussed above holds. That is, with ρ_1 complex, there is no unique y_t lying in $H_{\bar{x}}(t)$ which satisfies (14).

For the first order scalar expectational difference equation, there are either many solutions lying in $H_{\bar{x}}(t)$ or just one. But in the second order case, there arises the possibility that no solution lying in $H_{\bar{x}}(t)$ exists. The roots ρ_1^{-1} and ρ_2^{-1} of the characteristic equation of the model (14) completely determine which case arises. The second order case is the simplest situation in which all three possibilities exist. Since when there are many solutions in $H_{\bar{x}}(t)$ it is hard to imagine solutions in other spaces, the class of solutions might be characterized as: many, one, and no solutions in $H_{\bar{x}}(t)$ correspond to no, many, and all solutions to (14) elsewhere.

3. The nth Order Case

This section investigates the relationship between solutions in $H_{\bar{x}}(t)$, the number of "expectations terms" in the difference equation, and the number of characteristic roots inside the unit circle. The results of the previous section suggest $(1,1,1)$, $(0,1,2)$, and $(\infty,1,0)$ are possible. But what of higher order models? The equation to be studied is[3]

$$(21) \quad E_t\left(\sum_{j=-m}^{n} \lambda_j y_{t+j}\right) = x_t, \qquad \lambda_n = 1.$$

Such equations appear, for example, in the work on the term structure of interest rates by Shiller (1972), Sargent (1979a), and Hansen and Sargent (1981b).

---

[3]The information set at time t includes $\{y_s\}_{-\infty}^{t}$ and $\{x_s\}_{-\infty}^{t}$. Thus $E_t y_{t-r} = y_{t-r}$ for $r \geq 0$. This convention is also used in Chapter IV.

To study (21), assume first that $\{x_t\}$ has the Wold representation

(22) $\quad x_t = A(L)\varepsilon_t.$

Next, assume that $\{y_t\}$ can be written

(23) $\quad y_t = C(L)\varepsilon_t.$

Before proceeding to analyze (21), it will prove useful to calculate the terms $E_t y_{t+j}$ $j=1,\ldots,n$. Using (23) and the Wiener-Kolmogorov formulas,

$$E_t y_{t+j} = \left[\frac{C(L)}{L^j}\right]_+ \varepsilon_t$$

$$= L^{-j}(C(L) - \sum_{k=0}^{j-1} C_k L^k)\varepsilon_t.$$

Breaking (21) into two pieces and using the above expressions, the equation becomes

(24) $\quad \{\sum_{j=1}^{n} \lambda_j L^{-j}(C(L) - \sum_{k=0}^{j-1} C_k L^k)\}\varepsilon_t + \{\sum_{j=-m}^{0} \lambda_j L^{-j} C(L)\}\varepsilon_t = A(L)\varepsilon_t.$

Because (24) must hold for all realizations of $\{\varepsilon_t\}$, the z-transforms of the sequences represented in (24) must satisfy

$$\sum_{j=1}^{n} \lambda_j z^{-j}(C(z) - \sum_{k=0}^{j-1} C_k z^k) + \sum_{j=-m}^{0} \lambda_j z^{-j} C(z) = A(z).$$

This expression can be rearranged to yield

$$\sum_{j=-m}^{n} \lambda_j z^{-j} C(z) = A(z) + \sum_{j=1}^{n} \lambda_j z^{-j} \sum_{k=0}^{j-1} C_k z^k.$$

Define $\lambda(z) = \lambda_n + \lambda_{n-1}z + \ldots + \lambda_1 z^{n-1} + \lambda_0 z^n + \lambda_{-1} z^{n+1} + \ldots + \lambda_{-m} z^{n+m}.$

Upon multiplication by z^n, the above expression becomes

$$\lambda(z)C(z) = z^n A(z) + \sum_{j=1}^{n} \lambda_j \sum_{k=0}^{j-1} C_k z^{n-(j-k)},$$

or, after a change in the order of summation,

$$\lambda(z)C(z) = z^n A(z) + \sum_{k=0}^{n-1} C_k \sum_{j=k+1}^{n} \lambda_j z^{n-j+k}.$$

Next suppose that $\lambda(z)$ can be factored as

$$\lambda(z) = \prod_{j=1}^{n+m} (1-\rho_j z)$$

with $\rho_1, \ldots, \rho_{n+m}$ distinct, ordered as $|\rho_1| > |\rho_2| > \ldots > |\rho_{n+m}|$, and with $\rho_1, \ldots, \rho_r$ outside the unit circle. There are, again, three interesting cases: $r = n$, $r < n$, and $r > n$.

Formally, the z-transform $C(z)$ of the coefficients in the moving average representation for y_t can be written

(25) $$C(z) = \{\prod_{j=1}^{n+m} (1-\rho_j z)\}^{-1} \{z^n A(z) + \sum_{k=0}^{n-1} C_k \sum_{j=k+1}^{n} \lambda_j z^{n-j+k}\}.$$

Notice that there are n as yet undetermined parameters $(C_0, \ldots, C_{n-1})$ on the right-hand side of (25). In the event that $r = n$, the $C(z)$ in (25) has n isolated singularities inside the unit circle. As in previous sections, the intention here is to attempt to set $C_0, \ldots, C_{n-1}$ so as to make $C(z)$ analytic on $|z| < 1$. To do this, one must set the residues of $C(\cdot)$ at $\rho_1, \ldots, \rho_r$ to zero:

$$0 = \lim_{z \to \rho_q^{-1}} (1-\rho_q z)C(z)$$

$$= \{ \prod_{\substack{j=1 \\ j \neq q}}^{n+m} (1-\rho_j \rho_q^{-1}) \}^{-1} \{ \rho_q^{-n} A(\rho_q^{-1}) + \sum_{k=0}^{n-1} C_k \sum_{j=k+1}^{n} \lambda_j \rho_q^{-(n-j+k)} \}$$

for $q = 1, \ldots, n$, which requires

(26) $\quad -\rho_q^{-n} A(\rho_q^{-1}) = \sum_{k=0}^{n-1} C_k \sum_{j=k+1}^{n} \lambda_j \rho_q^{-(n-j+k)}, \qquad q = 1, \ldots, n.$

Expression (26) is a set of n linear equations in the n unknowns C_0, $C_1, \ldots, C_{n-1}$. In matrix notation, (26) can be written

(27)
$$\begin{bmatrix} \sum_{j=1}^{n} \lambda_j \rho_1^{-(n-j)} & \sum_{j=2}^{n} \lambda_j \rho_1^{-(n+1-j)} & \cdots & \lambda_n \rho_1^{-(n-1)} \\ & & \cdot & \\ & & \cdot & \\ & & \cdot & \\ \sum_{j=1}^{n} \lambda_j \rho_n^{-(n-j)} & \sum_{j=2}^{n} \lambda_j \rho_n^{-(n+1-j)} & \cdots & \lambda_n \rho_n^{-(n-1)} \end{bmatrix} \begin{bmatrix} C_0 \\ C_1 \\ \cdot \\ \cdot \\ \cdot \\ C_{n-1} \end{bmatrix} = - \begin{bmatrix} \rho_1^{-n} A(\rho_1^{-1}) \\ \cdot \\ \cdot \\ \cdot \\ \rho_n^{-n} A(\rho_n^{-1}) \end{bmatrix}.$$

The nonsingularity of the (nxn) matrix M_n on the left side of (27) can be established as follows. Note that since $\lambda_n \neq 0$, the last column of the matrix is a column of distinct numbers. The $(n-1)^{th}$ column is obtained from the nth by multiplying the jth element by ρ_j and adding to the result $\lambda_{n-1} \rho_j^{-(n-1)}$. Thus, the $(n-1)^{th}$ column is linearly independent of the nth. Such a relationship holds for any two adjacent columns. Because the ρ_j^s are distinct, the ith column of the matrix is not expressible as a linear combination of the columns to its right.[4] Thus the n columns of the matrix

---

[4]When the ρ_j^s are not distinct, the matrix in question is different, but a similar argument goes through.

are linearly independent; the matrix is nonsingular. The unique solution to (27) is then

$$\begin{bmatrix} \hat{c}_0 \\ \cdot \\ \cdot \\ \cdot \\ \hat{c}_{n-1} \end{bmatrix} = M_n^{-1} \begin{bmatrix} -\rho_1^{-n} A(\rho_1^{-1}) \\ \cdot \\ \cdot \\ \cdot \\ -\rho_n^{-n} A(\rho_1^{-1}) \end{bmatrix}$$

Using these values and expression (26), the unique solution $\{y_t\}$ to (21) which lies in $H_{\tilde{x}}(t)$ is given by

$$(28) \quad y_t = \{ \prod_{j=1}^{n+m} (1-\rho_j L) \}^{-1} \{ L^n A(L) + \sum_{k=0}^{n-1} \hat{c}_k \sum_{j=k+1}^{n} \lambda_j L^{n-j+k} \} \varepsilon_t$$

$$= \{ \prod_{j=1}^{n+m} (1-\rho_j L) \}^{-1} \{ L^n A(L) + \sum_{\ell=0}^{n} (\sum_{p=0}^{\ell} \hat{c}_p \lambda_{n-\ell+p}) L^\ell \} \varepsilon_t .$$

When $r < n$, the analogue to (27) is

$$(29) \quad \begin{bmatrix} \sum_{j=1}^{n} \lambda_j \rho_1^{-(n-j)} & \sum_{j=2}^{n} \lambda_j \rho_1^{-(n+1-j)} & \cdots & \lambda_n \rho_1^{-(n-1)} \\ & \cdot & & \\ & \cdot & & \\ & \cdot & & \\ \sum_{j=1}^{n} \lambda_j \rho_r^{-(n-j)} & \sum_{j=2}^{n} \lambda_j \rho_r^{-(n+1-j)} & \cdots & \lambda_n \rho_r^{-(n-1)} \end{bmatrix} \begin{bmatrix} c_0 \\ \cdot \\ \cdot \\ \cdot \\ c_{n-1} \end{bmatrix} = - \begin{bmatrix} \rho_1^{-n} A(\rho_1^{-1}) \\ \cdot \\ \cdot \\ \cdot \\ \rho_r^{-n} A(\rho_r^{-1}) \end{bmatrix}$$

or

$$(30) \quad M_r \begin{bmatrix} c_0 \\ \cdot \\ \cdot \\ \cdot \\ c_{n-1} \end{bmatrix} = - \begin{bmatrix} \rho_1^{-n} A(\rho_1^{-1}) \\ \cdot \\ \cdot \\ \cdot \\ \rho_r^{-n} A(\rho_r^{-1}) \end{bmatrix} .$$

The matrix M_r is (rxn) with rank $r < n$, and thus has an infinity of right-inverses $M_{r(R)}^{-1}$ (see Noble (1969, p. 135)), each generating a solution

$$\begin{bmatrix} \bar{C}_0 \\ \cdot \\ \cdot \\ \cdot \\ \bar{C}_{n-1} \end{bmatrix} = -M_{r(R)}^{-1} \begin{bmatrix} \rho_1^{-n} A(\rho_1^{-1}) \\ \cdot \\ \cdot \\ \cdot \\ \rho_r^{-n} A(\rho_r^{-1}) \end{bmatrix}$$

In this case, (21) does not uniquely determine the representation of $\{y_t\}$ in $H_{\tilde{x}}(t)$:

$$y_t = \{\prod_{j=1}^{n+m}(1-\rho_j L)\}^{-1} \{L^n A(L) + \sum_{\ell=0}^{n}(\sum_{p=0}^{\ell} \bar{C}_p \lambda_{n-\ell+p})L^\ell\}\varepsilon_t$$

where $(\bar{C}_0,\ldots,\bar{C}_{n-1})$ is any solution to (29).

When $r > n$, the analogue to the matrix M_r in (30) is (rxn) with rank $n < r$. In this case (30) has at most one solution. In practice, though, such equations almost never have solutions. The results of the analysis are summarized in the following lemma.

Lemma 1. Suppose

(21) $\quad E_t(\sum_{j=-m}^{n} \lambda_j y_{t+j}) = x_t \qquad \lambda_n = 1$

with $\{x_t\}$ a LRCSSP^0 with Wold representation $x_t = \sum_{j=0}^{\infty} A_j \varepsilon_{t-j} = A(L)\varepsilon_t$. The information set at time t is $\{\varepsilon_t, \varepsilon_{t-1},\ldots\}$. Write $\lambda(z) = z^n \sum_{j=-m}^{n} \lambda_j z^{-j}$

$= \prod_{j=1}^{n+m}(1-\rho_j z)$. Suppose that $\rho_1,\ldots,\rho_r$ are outside the unit circle. Then

a) If $r = n$, there is a unique y_t in $H_{\bar{x}}(t)$ which satisfies (21).

b) If $r < n$, there is an infinity of y_t's in $H_{\bar{x}}(t)$ satisfying (21).

c) If $r > n$, there is no y_t in $H_{\bar{x}}(t)$ which satisfies (21).

Lemma 1 completely describes the class of solutions to the general scalar expectational difference equation which lies in the space spanned by the driving process.

Rational expectations models are not always describable by equations of the same form as (21). But simple modifications of (21) do mirror most such models. The consequences of these modifications are studied in the next three sections.

4. Bivariate Representations

In many contexts, particularly estimation, it is useful to think of $\{x_t\}$ and $\{y_t\}$ as jointly covariance stationary processes. Such processes have bivariate Wold representations of the form

$$(31) \quad \begin{bmatrix} x_t \\ y_t \end{bmatrix} = \begin{bmatrix} A(L) & B(L) \\ C(L) & D(L) \end{bmatrix} \begin{bmatrix} \varepsilon_{1t} \\ \varepsilon_{2t} \end{bmatrix} \equiv G(L)\varepsilon_t$$

where now $\{\varepsilon_{1t}, \varepsilon_{2t}\}$ are jointly fundamental for $\{x_t, y_t\}$. For (31) to be a fundamental representation for $[x_t \; y_t]'$, $\det(G(z))$ must be analytic on $|z| < 1$ (see, e.g., Whittle (1963, p. 29)). If, in addition, $\{x_t\}$ and $\{y_t\}$ are to be regular, $A(z)$, $B(z)$, $C(z)$, and $D(z)$ must be individually analytic on $|z| < 1$. (See Rozanov (1967, p. 58).)

A simple example will illustrate how to derive the cross-equation rational expectations restrictions in a bivariate context. Suppose the model is given by (1):

$$(1) \quad E_t y_{t+1} - \rho y_t = x_t.$$

The Wiener-Kolmogorov formulas, applied to (31), yield

$$E_t y_{t+1} = L^{-1}[C(L) - C_0]\varepsilon_{1t} + L^{-1}[D(L) - D_0]\varepsilon_{2t}$$

Thus (1) becomes

(32) $\quad L^{-1}[C(L) - C_0]\varepsilon_{1t} + L^{-1}[D(L) - D_0]\varepsilon_{2t} - \rho C(L)\varepsilon_{1t} - \rho D(L)\varepsilon_{2t}$

$$= A(L)\varepsilon_{1t} + B(L)\varepsilon_{2t},$$

which must hold for all realizations of ε_{1t} and ε_{2t}. But then

$$L^{-1}[C(L) - C_0]\varepsilon_{1t} - \rho C(L)\varepsilon_{1t} = A(L)\varepsilon_{1t}$$

and

$$L^{-1}[D(L) - D_0]\varepsilon_{2t} - \rho D(L)\varepsilon_{2t} = B(L)\varepsilon_{2t}.$$

Thus the z-transforms of the sequences represented by $C(L)$ and $D(L)$ must satisfy $C(z) = (1-\rho z)^{-1}(zA(z) + C_0)$ and $D(z) = (1-\rho z)^{-1}(zB(z) + D_0)$. If $|\rho| > 1$, and regularity is assumed, $C(z) = (1-\rho z)^{-1}\{zA(z) - \rho^{-1}A(\rho^{-1})\}$ and $D(z) = (1-\rho z)^{-1}\{zB(z) - \rho^{-1}B(\rho^{-1})\}$. Then (31), restricted by (1), is

(33) $\quad \begin{bmatrix} x_t \\ y_t \end{bmatrix} = \begin{bmatrix} A(L) & B(L) \\ \dfrac{LA(L) - \rho^{-1}A(\rho^{-1})}{1 - \rho L} & \dfrac{LB(L) - \rho^{-1}B(\rho^{-1})}{1 - \rho L} \end{bmatrix} \begin{bmatrix} \varepsilon_{1t} \\ \varepsilon_{2t} \end{bmatrix}$

Notice that the elements in the second row of the matrix in (33) are obtained from the corresponding elements of the first row in precisely the same way. For instance, write $K(A(L)) = (1-\rho L)^{-1}\{LA(L) - \rho^{-1}A(\rho^{-1})\}$. Then $D(L) = K(B(L))$. This is a consequence of the requirement that the model hold for all realizations, not of the form of the model. The result is summarized in the following lemma.

Lemma 2. Suppose the assumptions of Lemma 1 are satisfied and that $r = n$. Then if the unique solution y_t lying in $H_{\bar{x}}(t)$ can be written as $y_t = K(A(L))\varepsilon_t$, the bivariate representation for the regular process $\{x_t\ y_t\}'$ is of the form

$$\begin{bmatrix} x_t \\ y_t \end{bmatrix} = \begin{bmatrix} A_1(L) & B_1(L) \\ K(A_1(L)) & K(B_1(L)) \end{bmatrix} \begin{bmatrix} \varepsilon_{1t} \\ \varepsilon_{2t} \end{bmatrix}$$

The upshot of this result is that when studying models of the form (21), one may as well pursue the univariate analysis of section 3. Given the univariate result, the bivariate representation of the system is easily obtained. There are, however, three caveats. First, the lemma is not useful for the "perturbed equations" to be studied in section 5 below. Second, a relaxation of the regularity condition raises a uniqueness problem which will be studied briefly.

Suppose again that the model is (1):

$$E_t y_{t+1} - \rho y_t = x_t.$$

Suppose also that $|\rho| > 1$, so that there is a unique solution y_t in $H_{\bar{x}}(t)$. For (31) to be the fundamental representation for $\{x_t\ y_t\}$, $\det(G(z))$ must be analytic on $|z| < 1$. Proceeding formally using the calculations made above, this means that

$$\det \begin{bmatrix} A(z) & B(z) \\ \dfrac{zA(z) + C_0}{1 - \rho z} & \dfrac{zB(z) + D_0}{1 - \rho z} \end{bmatrix}$$

must be analytic on $|z| < 1$. Thus

$$\frac{A(z)\{zB(z) - D_0\} - B(z)\{zA(z) - C_0\}}{1 - \rho z} = \frac{-D_0 A(z) + C_0 B(z)}{1 - \rho z}$$

must be analytic on $|z| < 1$. This function clearly has a singularity at ρ^{-1}; its residue there is $-D_0 A(\rho^{-1}) + C_0 B(\rho^{-1})$. One way to set this to zero is to require $C_0 = -\rho^{-1} A(\rho^{-1})$ and $D_0 = -\rho^{-1} B(\rho^{-1})$. For these settings of C_0 and D_0, (33) results. But there are many other ways to make $\det(G(z))$ appropriately analytic. Thus, for any finite C_0,

$$\begin{bmatrix} x_t \\ y_t \end{bmatrix} = \begin{bmatrix} A(L) & B(L) \\ \dfrac{LA(L) + C_0}{1 - \rho L} & \dfrac{LB(L) + C_0 B(\rho^{-1}) A(\rho^{-1})^{-1}}{1 - \rho L} \end{bmatrix} \begin{bmatrix} \varepsilon_{1t} \\ \varepsilon_{2t} \end{bmatrix}$$

is a fundamental representation for an $\{x_t \ y_t\}'$ process which satisfies (21). Evidently, the regularity assumption eliminates this nonuniqueness.

The third point is that like the regularity assumption, the assumption that (31) is a fundamental representation is a powerful one. This point is easy to see in the present context. Returning to the unique solution represented by (33), notice that if $B(L) = 0$, then $D(L) = 0$. By Sims's (1972) theorem 1, $B(L) = 0$ in a moving average representation not restricted by a model like (1) is equivalent to the statement "y fails to Granger (1969)-cause x." But setting $B(L) = 0$ in (33) results in the univariate representation analyzed in section 1. In this case, x and y span the same space; neither variable Granger-causes the other. But in (33), x and y do not span the same space and each Granger-causes the other. This is a result of the assumption that the $(x_t \ y_t)$ process described by (31) is nonsingular. The singularity or nonsingularity of this process resurfaces in the next section, which investigates conditions under which one-way causal orderings are possible.

5. Perturbed Equations

Many of the rational expectations models which have appeared in the literature conform to what Hansen and Sargent (1981b) call "inexact models." That is, the rational expectation restrictions are perturbed by "noise" with a well-defined structure. For example, a model of this type is used in the Saracoglu and Sargent (1978) analysis of Cagan's portfolio balance schedule. Their version of the schedule is

$$m_t - p_t = \alpha(E_t p_{t+1} - p_t) + \eta_t$$

where $\{\eta_t\}$ is assumed to be strictly econometrically exogenous with respect to $\{m_t\}$: $Em_t \eta_{t-s} = 0$ ∀s. Another example is given by the necessary condition for an optimum in the factor demand analysis of Hansen and Sargent (1980a). Their equation, appropriately relabeled, is

$$E_t y_{t+1} + \phi\beta^{-1} y_t + \beta^{-1} y_{t-1} = \delta^{-1}(x_t - a_t),$$

where β is a discount factor, ϕ and δ are functions of parameters in the agent's objective function, and the process $\{a_t\}$ is not Granger-caused by either $\{y_t\}$ or $\{x_t\}$.

The simple equation (1) serves as a laboratory in which to study "perturbed equations." Modifying (1) slightly, the equation to be studied is

(34) $$E_t y_{t+1} - \rho y_t = x_t + \eta_t.$$

A number of interesting assumptions about $\{\eta_t\}$ are studied in a general context below.

Suppose $\{x_t\}$ and $\{y_t\}$ are jointly stationary and regular with Wold representation

(31) $$\begin{bmatrix} x_t \\ y_t \end{bmatrix} = \begin{bmatrix} A(L) & B(L) \\ C(L) & D(L) \end{bmatrix} \begin{bmatrix} \varepsilon_{1t} \\ \varepsilon_{2t} \end{bmatrix} \equiv G(L)\varepsilon_t.$$

Next, suppose that η_t is related to the innovation in $\{x_t\}$ by $\eta_t = a(L)\varepsilon_{1t} + b(L)\varepsilon_{2t}$. Finally, suppose that the information set implicit in (34) is $H_{\bar{\varepsilon}}(t)$. The interesting solutions y_t lie in this space. Using the techniques of the previous section, (34) may be written as

(35) $$L^{-1}[C(L) - C_0]\varepsilon_{1t} + L^{-1}[D(L) - D_0]\varepsilon_{2t} - \rho\{C(L)\varepsilon_{1t} + D(L)\varepsilon_{2t}\}$$
$$= A(L)\varepsilon_{1t} + B(L)\varepsilon_{2t} + a(L)\varepsilon_{1t} + b(L)\varepsilon_{2t}.$$

Expression (35) must hold for all realizations of $\{\varepsilon_{1t}\ \varepsilon_{2t}\}$, so the z-transforms of the sequences represented there must satisfy $C(z) = (1-\rho z)^{-1}\{zA(z) + za(z) + C_0\}$ and $D(z) = (1-\rho z)^{-1}\{zB(z) + zb(z) + D_0\}$. When $|\rho| > 1$, the regular process $\{x_t\ y_t\}$ can be represented as

(36) $$\begin{bmatrix} x_t \\ y_t \end{bmatrix}$$
$$= \begin{bmatrix} A(L) & B(L) \\ \dfrac{LA(L)-\rho^{-1}A(\rho^{-1})}{1-\rho L} - \dfrac{La(L)-\rho^{-1}a(\rho^{-1})}{1-\rho L} & \dfrac{LB(L)-\rho^{-1}B(\rho^{-1})}{1-\rho L} - \dfrac{Lb(L)-\rho^{-1}b(\rho^{-1})}{1-\rho L} \end{bmatrix}$$
$$\times \begin{bmatrix} \varepsilon_{1t} \\ \varepsilon_{2t} \end{bmatrix}.$$

Notice that the introduction of η_t has made lemma 2 inapplicable.

Under alternative assumptions on a(L) and b(L), (36) can produce each of the three possible Granger orderings. First, suppose a(L) = b(L) ≡ 0. This case was analyzed in the previous section: x and y Granger-cause each other. Second, suppose A(L) = b(L) ≡ 0. Then (36) becomes

$$(37) \quad \begin{bmatrix} x_t \\ y_t \end{bmatrix} = \begin{bmatrix} 0 & B(L) \\ \dfrac{\rho^{-1}a(\rho^{-1}) - La(L)}{1 - \rho L} & \dfrac{LB(L) - \rho^{-1}B(\rho^{-1})}{1 - \rho L} \end{bmatrix} \begin{bmatrix} \varepsilon_{1t} \\ \varepsilon_{2t} \end{bmatrix}.$$

By Sims's theorem 1, y fails to Granger-cause x. In this case, y_t does not lie in $H_{\bar{x}}(t)$, though it does lie in $H_{\bar{\varepsilon}}(t)$. Finally, suppose a(L) = -A(L), b(L) ≡ 0. Then (36) becomes

$$\begin{bmatrix} x_t \\ y_t \end{bmatrix} = \begin{bmatrix} A(L) & B(L) \\ 0 & \dfrac{LB(L) - \rho^{-1}B(\rho^{-1})}{1 - \rho L} \end{bmatrix} \begin{bmatrix} \varepsilon_{1t} \\ \varepsilon_{2t} \end{bmatrix},$$

in which case x fails to Granger-cause y. Thus y lies in a proper subspace of $H_{\bar{x}}(t)$.

The second example is perhaps the most important. If x has an autoregressive representation, then by Sims's (1972) theorem 2, y can be expressed as a distributed lag on current and past x with a strictly exogenous residual. The cross-equation restrictions remain: the coefficients in the distributed lag are functions of those in the autoregressive representation for x. Thus, the relation between y and x is not exact, as it is in many rational expectations models.

The moving average representation (37) is qualitatively similar to the one in Hansen and Sargent (1980a). Their error term arises because agents possess more information than econometricians--an assumption which guarantees that the multivariate process they study is nonsingular. But it is clear from this and the preceding section that such an assumption concerning information sets is not necessary. That is, the multivariate process may be assumed to be nonsingular before restrictions like (1) or (34) are imposed. The question of why the system is (or is not) nonsingular remains; the point is that error terms in y on x regressions computed from restricted representations like (37) can be present even if agents and econometricians possess the same information.

6. Withholding Equations

A withholding equation is a special type of expectational difference equation in which some relevant information is concealed from agents. Operationally, this means that information sets do not contain values for some of the variables in the difference equation. The simplest example of such an equation is a slight modification of (1):

(38) $\quad E_{t-1} y_t - \rho y_t = x_t$.

This equation was obtained by applying the lag operator L to the first term in (1)[5].

Versions of (38) are, perhaps, more prevalent in the literature than versions of (1). A whole class of models stemming from the absolute-, relative-price confusion paradigm of Lucas (1973, 1975) employs (38). This class includes, for example, the model used by Sargent and Wallace (1975, 1976) to study conventional macroeconomic policy design. The model consisted of two equations,

---

[5] (1) was $E_t y_{t+1} - \rho y_t = x_t$. Applying L, $L E_t y_{t+1} - \rho y_t = x_t$ becomes (38).

$$y_t = \alpha(p_t - E_{t-1}p_t) + \lambda y_{t-1} + u_t$$

$$m_t - p_t = y_t + \eta_t.$$

The variables m_t and p_t are as above, y_t is the log of real GNP, and u_t and η_t are serially uncorrelated zero-mean stochastic processes. The first equation is a Lucas (1972b)-type Phillips curve; only unexpected movements in prices affect output. The second equation is a simple portfolio balance schedule. By solving this latter equation for y_t and substituting it into the first equation, one obtains the second-order equation

(39) $\quad E_{t-1}p_t - ((1+\alpha)/\alpha)p_t + \lambda p_{t-1} = u_t - (1-\lambda L)(m_t - \eta_t).$

Equation (39) is a withholding equation because both $E_{t-1}p_t$ and p_t appear.

An additional reason for the prevalence of withholding equations in the literature is that Muth's (1961) first example of a rational expectations model produced such an equation. Muth set out to study price fluctuations in an isolated market by positing the model

$$c_t = -\beta p_t \quad \text{(Demand)}$$

$$p_t = \gamma E_{t-1}p_t + u_t \quad \text{(Supply)}$$

$$p_t = c_t \quad \text{(Market equilibrium)}$$

where c_t is consumption at t, p_t is the market price at t, and u_t is a white noise error term. By substituting the first and third equations into the second, one obtains the first order withholding equation

(40) $\quad E_{t-1}p_t + (\beta/\gamma)p_t = -\gamma^{-1}u_t.$

Withholding equations have been studied by Aoki and Canzoneri (1979). The analysis of the next several pages pursues the link their work began to draw between withholding equations and the expectational difference equations of the previous five sections.

To study withholding equations, it will prove useful to abstract from economic interpretations and return to (38):

(38) $\quad E_{t-1} y_t - \rho y_t = x_t.$

As in previous sections, it will be assumed that $\{x_t\}$ possesses the Wold representation $x_t = A(L)\varepsilon_t$, and solutions to (38) of the form $y_t = C(L)\varepsilon_t$ will be sought. The first chore is to calculate $E_{t-1} y_t$. Using the fact that $y_t = C(L)\varepsilon_t$,

$$\begin{aligned} E_{t-1} y_t &= E_{t-1}[C_0 \varepsilon_t + C_1 \varepsilon_{t-1} + C_2 \varepsilon_{t-2} + \ldots] \\ &= C_1 \varepsilon_{t-1} + C_2 \varepsilon_{t-2} + \ldots \\ &= L[C_1 + C_2 L + C_3 L^2 + \ldots]\varepsilon_t \\ &= [C(L) - C_0]\varepsilon_t. \end{aligned}$$

Then (38) becomes

$$[C(L) - C_0]\varepsilon_t - \rho C(L)\varepsilon_t = A(L)\varepsilon_t$$

which must hold for all realizations of $\{\varepsilon_t\}$. Thus, the solution can be calculated using the z-transform:

$$C(z) - C_0 - \rho C(z) = A(z)$$

or, formally,

(41) $\quad C(z) = \dfrac{A(z) + C_0}{1 - \rho}.$

For any finite C_0, the function $C(z)$ is clearly analytic on the open unit disk. Superficially, then, (41) appears to embody a uniqueness problem. However, (41) is different from the formal solutions of previous sections in an important way. In those cases, evaluation of $C(z)$ at $z = 0$ resulted in the trivial equality $C_0 = C_0$. But using (41),

$$C_0 = \frac{A(0) + C_0}{1 - \rho}.$$

Consistency requires that $C_0 = -\rho^{-1}A(0)$. In this case, $C(z)$ is the z-transform of a square-summable sequence, and

(42) $$y_t = \frac{A(L) - \rho^{-1}A(0)}{1 - \rho}\varepsilon_t.$$

Equation (42) decribes the unique solution y_t lying in $H_{\bar{x}}(t)$ which satisfies (38). It is important to note that such a unique solution exists <u>regardless of the magnitude of ρ</u>. The reason for this is that the lagged expectation term in (38) causes the characteristic equation to have a root at zero rather than at ρ^{-1}. To see this, write $C(L) = \gamma + L\Gamma(L)$, with $\Gamma(L) = \sum_{j=0}^{\infty} \Gamma_j L^j$. Then $E_{t-1}y_t = L(E_t y_{t+1}) = L(\Gamma(L)\varepsilon_t)$. Then the z-transform of the coefficients implicit in (38) is

$$z\Gamma(z) - \rho(\gamma + z\Gamma(z)) = A(z)$$

or

$$(1-\rho)z\Gamma(z) = A(z) + \rho\gamma.$$

Then

$$\Gamma(z) = \frac{A(z) + \rho\gamma}{(1 - \rho)z}$$

has an isolated singularity at $z = 0$. To remove this singularity, set γ to make the residue of $\Gamma(z)$ at $z = 0$ equal to zero:

$$\lim_{z \to 0} z\Gamma(z) = 0$$

$$= A(0) + \rho\gamma,$$

or $\gamma = -\rho^{-1}A(0)$, as above.

Evidently, the conditions for existence and uniqueness of solutions to withholding equations are quite different from those for the general expectational difference equation. Indeed, consider

(43) $\quad E_{t-2}y_t - \rho y_t = x_t.$

Using $x_t = A(L)\varepsilon_t$ and $y_t = C(L)\varepsilon_t$, the z-transform of the coefficients in (43) becomes

$$C(z) - C_0 - C_1 z - \rho C(z) = A(z)$$

or

$$C(z) = \frac{A(z) + C_0 + C_1 z}{1 - \rho}.$$

Evaluating this expression at $z = 0$,

$$C_0 = \frac{A(0) + C_0}{1 - \rho}$$

and $C_0 = \rho^{-1}A(0)$ as above. Thus

$$C(z) = \frac{A(z) - \rho^{-1}A(0) + C_1 z}{1 - \rho}.$$

Calculating the derivative of $C(z)$ and evaluating the result at $z = 0$, one obtains

$$C_1 = \frac{A'(0) + C_1}{1 - \rho}$$

which requires $C_1 = \rho^{-1}A'(0)$. Thus, the unique solution y_t to (43) which lies in $H_{\bar{x}}(t)$ is given by

$$y_t = (1-\rho)^{-1}\{A(L) - \rho^{-1}A(0) - \rho^{-1}LA'(0)\}\varepsilon_t.$$

This result is summarized in lemma 3.

Lemma 3. Suppose

$$E_{t-n}y_t - \rho y_t = x_t$$

where the Wold representation of the LRCSSP^0 $\{x_t\}$ is $x_t = A(L)\varepsilon_t$. Then the unique solution y_t lying in $H_{\bar{x}}(t)$ is

$$y_t = (1-\rho)^{-1}\{A(L) - \rho^{-1}\sum_{j=0}^{n-1} j_A(j)(0)\}$$

where $A^{(j)}(0) = \frac{d^j}{dz^j}A(z)|_{z=0}$.

This lemma appears to suggest that the coefficients of the withholding equation do not affect the uniqueness of the solution. The suggestion is false. Consider the equation

(44) $\quad E_{t-1}y_t + \rho_1 y_t + \rho_2 y_{t-1} = x_t.$

Formally, the solution is represented by

(45) $\quad C(z) = \frac{A(z) + C_0}{1 + \rho_1 + \rho_2 z}.$

Suppose first that $|1 + \rho_1| > |\rho_2|$. Then $C(z)$ is analytic on the open unit disk and, as above, $C_0 = -\rho_1^{-1}A(0)$. Then

(46) $\quad y_t = (1+\rho_1+\rho_2 L)^{-1}\{A(L) - \rho_1^{-1}A(0)\}\varepsilon_t$

is the only solution lying in $H_{\bar{x}}(t)$. Now suppose that $|1 + \rho_1| < |\rho_2|$. Then (45) cannot be both consistent and analytic: consistency requires that $C_0 = -\rho_1^{-1}A(0)$, while analyticity requires that

$$\lim_{z \to \frac{1+\rho_1}{\rho_2}} (1+\rho_1+\rho_2 z)C(z) = 0$$

$$= A\left(\frac{1+\rho_1}{\rho_2}\right) + C_0$$

or $C_0 = -A((1+\rho_1)/\rho_2)$. In this case no solution y_t in $H_{\bar{x}}(t)$ exists. The reason for this is that, as in section 2, there are <u>two</u> roots inside the unit circle: one at zero, the other at $(1+\rho_1)/\rho_2$.

The example of the previous paragraph illustrates two facts. First, the conditions under which withholding equations fail to have solutions are weaker than those for general expectational difference equations. Second, although lemma 3 covers the withholding equations studied by Aoki and Canzoneri (1979), it does not exhaust the class of interesting cases. To see this, consider

(47) $\quad E_{t-s} \sum_{j=-u}^{n} \lambda_j y_{t+j} + \sum_{j=-m}^{0} \beta_j y_{t+j} = x_t$

where $x_t = A(L)\varepsilon_t$. The case $s = 0$, $u \geq 0$ was dealt with in lemma 1. The interesting case, which covers the above example, is $u \geq s > 0$. Using $y_t = C(L)\varepsilon_t$ and proceeding as for lemma 1, one obtains

$$\left(\sum_{j=-u}^{n} \lambda_j z^{-j} + \sum_{j=-m}^{0} \beta_j z^{-j}\right) C(z) = A(z) + \sum_{j=-u}^{n} \lambda_j z^{-j} \sum_{k=0}^{j+s-1} C_k z^k.$$

Define $\tilde{\lambda}(z) = \sum_{j=-u}^{n} \lambda_j z^{n-j} + \sum_{j=-m}^{0} \beta_j z^{n-j}$. Then

(48) $\quad \tilde{\lambda}(z) C(z) = z^n A(z) + \sum_{j=-u}^{n} \lambda_j \sum_{k=0}^{j+s-1} C_k z^{n-(j-k)}$

where the order of the polynomial $\tilde{\lambda}(z)$ is $n + \max(u,m)$. Suppose that $\tilde{\lambda}(z)$ can be factored as $\tilde{\lambda}(z) = \prod_{q=1}^{n+\max(u,m)} (1-\tilde{\rho}_q z)$, and that $\rho_1 > \rho_2 > \rho_3 > \ldots$ with ρ_r the smallest factor satisfying $|\rho_q| > 1$. Since there are $n + s$ as yet undetermined coefficients in (48), it is clear that if $r > n + s$, no solution y_t lying in $H_x^-(t)$ exists. Expression (44) indicates that when $r = n + s$, a solution need not exist. In that case $n = u = 0$, $m = 1 = s$, and r was equal to 1, giving $r = 1 = n + s$. Since each of the four parameters s, u, n, and m impinges on the nature of the solution, a general theorem is not likely to be useful: applications will generally possess simple configurations of those parameters. Thus, a solution search which relies on the direct application of the z-transform methods to the problem at hand will probably be simpler than reference to a general theorem.

Conclusion

Simple variations of the methods presented in this chapter can be used to solve a large class of rational expectations models. The methods, unlike most "undetermined coefficients" techniques, are quite mechanical and require little guessing. The application of these techniques to current problems in rational expectations modeling, a comparison to standard methods, and extensions to multivariate models are discussed in the next several chapters.

II

Solution Techniques for Rational Expectations Models: A Critical Review

If one reads between the lines of the recent rational expectations literature, one can find there an implicit controversy awaiting transformation into an explicit one. The unstated issue concerns the efficacy of various algorithms for finding reduced forms in rational expectations models. Such models are, for the most part, systems of linear stochastic difference equations in which agents' views about the future enter in a fundamental way. If these expectations are formed adaptively, using fixed linear functions of available information, the problem of finding a reduced form is equivalent to finding the solution to a standard difference equation. The techniques for finding such solutions are well known.

But when expectations are assumed to be rational, the search for a reduced form is transformed into a rather more complicated search for a fixed point. Muth's (1961) original definition of the concept makes this clear; he defined rational expectations as "the predictions of the relevant economic theory." Thus suppose that expectations of future variables of interest can be written as linear functions of available information. Using these functions, one can calculate the reduced form of the model using the standard techniques. Then the reduced form can be used to forecast the same variables that agents were interested in. The model is solved when the forecasts of agents match those of the reduced form.

To find reduced forms with the fixed point property of rational expectations, it is necessary to employ special techniques. Though the equations to be solved are still difference equations, they are special ones: "expectational difference equations." To the casual reader of the rational expectations literature, it might appear that each author has his own solution technique. Though this taxonomy is far from being a partition, one can find "state-space" techniques in Lucas (1972a), operator methods in Sargent (1979b) and Wallis (1980), "methods of undetermined coefficients" using autoregressions and moving averages in Muth (1961) and Aoki and Canzoneri (1979), "forward" and "backward" solutions in Blanchard (1979), and most of the possible permutations of these techniques. In fact, this batch of techniques can be distilled to two distinct methods: one transforms an expectational difference equation into another, possibly simpler, ordinary difference equation; the other transforms the expectational difference equation into a system of nonlinear algebraic equations.

In addition to these techniques, there is another recently developed technique for finding solutions. The technique, a "method of undetermined coefficients in the frequency domain," was introduced by Saracoglu and Sargent (1978), simplified by Futia (1981), and extended and applied in a variety of situations in Chapter I. The simplicity and broad applicability of this technique, in comparison to the standard techniques, suggests that the time is ripe for making an issue of the choice of solution algorithms.

To make the issue explicit, it will be argued here that in comparison to the frequency domain undetermined coefficients methods, the standard techniques go too far toward obtaining closed form solutions, are too often intractable, or conceal the existence and uniqueness problems which often arise in rational expectations models. The comparisons proceed as follows. Section 1 below lays out a pair of simple expectational difference equations. As a standard by which to judge the others, the technique of Chapter I is then used to solve these equations. The next four sections analyze in turn each of four extant solution techniques in light of the section 1 solutions. Finally, because it is of independent interest, the Saracoglu-Sargent (1978) contraction mapping/undetermined coefficients method is discussed in section 6. It will be apparent that some of the details omitted from their paper are at the heart of the recent developments of this technique--developments which simply make the frequency domain undetermined coefficients method easier to use.

1. A Benchmark Solution Technique

Many, if not most, rational expectations models can be considered as particular incarnations of the simple first order expectational difference equation

(1) $\quad E_t y_{t+1} - \rho_1 y_t = x_t$

or of the second order equation

(2) $\quad E_t y_{t+1} - (\rho_1 + \rho_2) y_t + \rho_1 \rho_2 y_{t-1} = x_t.$

For most purposes, t indexes the integers, x_t and y_t are representatives of scalar stochastic processes, ρ_1 and ρ_2 are real numbers, and "$E_t y_{t+1}$" means "the best linear forecast of y_{t+1} based on information available at time t." When a model is characterized by (1) or (2), its reduced form is simply the solution of the equation which expresses y_t as a function of current and past values of x_t -- the information available at time t. When $\{x_t\}$ is a linearly regular covariance stationary stochastic process, the interesting class of solutions is the one for which $\{y_t\}$ is also regular and covariance stationary. Methods for finding this class of solutions, described in detail in Chapter I, will be reviewed briefly here.

Suppose $\{x_t\}$ has the given Wold representation

(3) $$x_t = \sum_{j=0}^{\infty} A_j \varepsilon_{t-j}$$

so that $\varepsilon_t = x_t - E(x_t | x_{t-1}, x_{t-2}, \ldots)$, $\sum_{j=0}^{\infty} A_j^2 < \infty$, and from Whittle (1963), the function $A(z) = \sum_{j=0}^{\infty} A_j z^j$ is analytic on the open unit disk. The representation (3) will often be written $x_t = A(L)\varepsilon_t$, where L is the lag operator: $L^n \varepsilon_t = \varepsilon_{t-n}$. Suppose further that agents know the linear structure of the x process, equation (3), and that they know its entire history, $\{x_s\}_{-\infty}^{t}$. Then the solution $\{y_t\}$ to (1) or (2) will depend on this information, and it is natural to write

(4) $$y_t = \sum_{j=0}^{\infty} C_j \varepsilon_{t-j} = C(L)\varepsilon_t$$

where the coefficients $C_0, C_1, \ldots$ are to be determined. If $\{y_t\}$ is to be stationary, $\operatorname{var} y_t = \sigma_\varepsilon^2 \sum_{j=0}^{\infty} C_j^2$ must be finite or, as above, $C(z)$ must be analytic on the open unit disk. From (4), $E_t y_{t+1}$ can be calculated using the Wiener-Kolmogorov formula[1]:

$$E_t y_{t+1} = \left[\frac{C(L)}{L} \right]_+ \varepsilon_t$$

---

[1] The annihilator operator $[\ ]_+$ means "ignore negative powers of L."

or

(5) $$E_t y_{t+1} = L^{-1}(C(L) - C_0)\varepsilon_t.$$

To find a solution to the first order equation, substitute (3), (4), and (5) into (1):

$$L^{-1}(C(L) - C_0)\varepsilon_t - \rho_1 C(L)\varepsilon_t = A(L)\varepsilon_t.$$

Since this expression must hold for any realization of $\{\varepsilon_t\}$, the sequences represented by the polynomials in the lag operator must be equal. Then, by the Riesz-Fischer theorem, the complex functions $z^{-1}(C(z) - C_0) - \rho_1 C(z)$ and $A(z)$ must be identical. Thus

(6) $$C(z) = (1-\rho_1 z)^{-1}(zA(z) + C_0).$$

If $|\rho_1| < 1$, then (6) represents an analytic function whenever C_0 is finite. Thus the solution to (1) is

(7) $$y_t = (1-\rho_1 L)^{-1}(LA(L)+C_0)\varepsilon_t = C(L)\varepsilon_t.$$

But when $|\rho_1| > 1$, $C(z)$ has an isolated singularity at ρ_1^{-1}. However, C_0 can be set in such a way as to make $C(z)$ analytic at ρ_1^{-1}. What is required is that C_0 be set to remove the residue of $C(z)$ at ρ_1^{-1}:

$$\lim_{z \to \rho_1^{-1}} (1-\rho_1 z)C(z) = 0 = \rho_1^{-1}A(\rho_1^{-1}) + C_0,$$

which requires $C_0 = -\rho_1^{-1}A(\rho_1^{-1})$. Then the (obviously unique) stationary solution to (1) can be written

(8) $$y_t = (1-\rho_1 L)^{-1}(LA(L) - \rho_1^{-1}A(\rho_1^{-1}))\varepsilon_t.$$

Clearly, (1) has a stationary solution regardless of the magnitude of ρ_1, though the solution is unique only for $|\rho_1| > 1$. But (2) need not have a stationary solution. This can be seen by using (3), (4), and (5) in (2) to obtain an expression analogous to (6):

(9) $\qquad C(z) = \{(1-\rho_1 z)(1-\rho_2 z)\}^{-1}\{zA(z) + C_0\}.$

If $|\rho_1| < 1$ and $|\rho_2| < 1$, the function given in (9) is analytic on the open unit disk for any finite C_0, and the solution to (2) is given by the natural analogue to (7). When $|\rho_1| < 1$ and $|\rho_2| > 1$, the $C(z)$ in (9) has an isolated singularity at ρ_2^{-1} which can be removed by setting $C_0 = -\rho_2^{-1}A(\rho_2^{-1})$. In this case, the solution to (2) is given by an equation similar to (8). But when $|\rho_1| > 1$ and $|\rho_2| > 1$, $C(z)$ has two isolated singularities inside the unit circle, and C_0 can be set to eliminate only one of them. Thus, in this case, $\{y_t\}$ cannot be written as a square-summable linear combination of current and past values of ε_t (or x_t). Under the natural interpretation of the model as causal with x as input and y as output, then, there is no stationary solution to (2).

If x_t has an autoregressive representation, the solutions (7), (8), or their second order analogues, can be converted into expressions giving y_t as the sum of current and past values of x_t, provided ε_t is replaced by $A(L)^{-1}x_t$. Then, for instance, the solution to (1) with $|\rho_1| > 1$ is given by

(10) $\qquad y_t = (1-\rho_1 L)^{-1}(L-\rho_1^{-1}A(\rho_1^{-1})A(L)^{-1})x_t.$

Expression (10) represents a solution of the form

(11) $\qquad y_t = \sum_{j=0}^{\infty} d_j x_{t-j}.$

But (10) does not give the sequence $\{d_j\}$ directly; rather, it embodies a generating function for d_j, leaving complicated convolutions unreduced. While $\{C_j\}$ and $\{d_j\}$ can be calculated by using (6) or (9) and the techniques for inverting z-transforms discussed, for instance, by Gabel and Roberts (1973), it seems sufficient to leave the solution in the form of (8) or (10), with the knowledge that, for instance, d_j is the coefficient on L^j in $(1-\rho_1 L)^{-1}(L-\rho_1^{-1} A(\rho_1^{-1}))A(L)^{-1}) = C(L)A(L)^{-1}$. The reason for this is that virtually all of the interesting information about the joint moments of the processes $\{y_t\}$ and $\{x_t\}$ can be obtained using (10) or, for that matter, the functions $C(z)$ and $A(z)$. For instance, from Sargent (1979b, Chapter XI), the covariance generating functions for y and x are $g_y(z) = \sigma^2 C(z)C(z^{-1})$ and $g_x(z) = \sigma^2 A(z)A(z^{-1})$, while the cross-covariance generating function is $g_{yx}(z) = \sigma^2 C(z)A(z^{-1})$. The coefficients on z^τ in these functions are $Ey_t y_{t-\tau}$, $Ex_t x_{t-\tau}$, and $Ey_t x_{t-\tau}$. The spectra and cross-spectrum can be obtained by evaluating these functions at $z = e^{-iw}$. Indeed, it is the identification of w with "frequency," spectra with "frequency domain," and the use of analyticity to determine C_0 which renders the z-transform solution technique discussed above a "method of undetermined coefficients in the frequency domain." Since $d(z) = g_{yx}(z)/g_x(z)$, the sequence $\{d_j\}$ and all of the variances and covariances can be obtained by inverse Fourier transforming $g_{yx}(e^{-iw})/g_x(e^{-iw})$, $g_{yx}(e^{-iw})$, $g_y(e^{-iw})$, and $g_x(e^{-iw})$. In addition, an easily obtained bivariate version of the solution (Chapter I, section 4) can be estimated by maximum likelihood methods using the Hannan (1970) spectral approximation outlined in Phadke and Kedem (1978), Kohn (1979), and Hansen and Sargent (1980a).

Clearly, a solution technique which provides for the calculation of $C(z)$ is quite powerful. Of course, methods which give closed form expressions for $\{d_j\}$ and $\{C_j\}$ are also illuminating. But such techniques are often quite complicated, and sometimes conceal the existence and uniqueness issues made transparent by the method used above. Substantiation for this claim is pursued in the next several sections.

2. Time Domain: Moving Averages

When John Muth (1961) first introduced the notion of rational expectations, he solved a simple model with a method which, for reasons which will become obvious, can be termed a "method of undetermined coefficients in the time domain using moving averages." Aoki and Canzoneri (1979) have recently generalized Muth's procedure. Though this method is similar in spirit to the method of section 1, there is one crucial difference: Muth's technique does not make use of the equivalence of the space of square-summable sequences and the space of analytic functions on the open unit disk--the equivalence, essentially, of the Riesz-Fischer theorem.

Muth's technique is easily demonstrated in the context of equation (1). First, substitute (3) and (4) into (1) to obtain

$$E_t \left(\sum_{j=0}^{\infty} C_j \varepsilon_{t+1-j} \right) - \rho_1 \sum_{j=0}^{\infty} C_j \varepsilon_{t-j} = \sum_{j=0}^{\infty} A_j \varepsilon_{t-j}.$$

Since $E_t \varepsilon_{t+1} = 0$, this expression can be written

$$\sum_{j=0}^{\infty} C_{j+1} \varepsilon_{t-j} - \rho_1 \sum_{j=0}^{\infty} C_j \varepsilon_{t-j} = \sum_{j=0}^{\infty} A_j \varepsilon_{t-j},$$

which must hold for all realizations of $\{\varepsilon_t\}$. Thus the coefficient on ε_{t-s} on the left-hand side of this expression must equal A_s for all s, or $C_1 - \rho_1 C_0 = A_0$, $C_2 - \rho_1 C_1 = A_1$, etc. Therefore, Muth's technique converts the expectational difference equation (1) into the ordinary difference equation

(12) $C_j - \rho_1 C_{j-1} = (1-\rho_1 L)C_j = A_{j-1}$ $j = 1, 2, \ldots$.

The general solution to this equation is obtained using standard techniques:

$$C_j = (1-\rho_1 L)^{-1} A_{j-1} + k\rho_1^j$$

where k is a constant to be determined.

Following Sargent (1979b, Chapter IX), C_j can be represented either as

(13a) $$C_j = \sum_{s=0}^{\infty} \rho_1^s A_{j-1-s} + k_1 \rho_1^j$$

or as

(13b) $$C_j = -\rho_1^{-1} \sum_{s=0}^{\infty} \rho_1^{-s} A_{j+s} + k_2 \rho_1^j$$

where A_s is understood to be 0 for $s < 0$. Because $\sum_{j=0}^{\infty} A_j^2 < \infty$, the infinite series in (13a) will generally not exist for all j when $|\rho_1| > 1$ while the one in (13b) will generally not exist when $|\rho_1| < 1$. In cases when both exist, the boundary condition will force k_1 and k_2 to be chosen to make (13a) and (13b) identical. Thus suppose $|\rho_1| < 1$ and consider (13a). The requirement that $\sum_{j=0}^{\infty} C_j^2 < \infty$ does not impose any condition on k_1, and C_j is of exponential order less than unity. Thus the solution can be written

(14) $$y_t = \sum_{j=0}^{\infty} (\sum_{s=0}^{\infty} \rho_1^s A_{j-1-s} + k_1 \rho_1^j) \varepsilon_{t-j}$$

which is identical to (7), with $k_1 = C_0$. When $|\rho_1| > 1$, (13b) is appropriate. In this case, k_2 must be set to zero to make $\sum_{j=0}^{\infty} C_j^2 < \infty$. Thus, as in (8),

$$C_0 = -\rho_1^{-1} \sum_{s=0}^{\infty} \rho_1^{-s} A_s = -\rho_1^{-1} A(\rho_1^{-1}),$$

and the solution is

(15) $$y_t = \sum_{j=0}^{\infty} (-\rho_1^{-1} \sum_{s=0}^{\infty} \rho_1^{-s} A_{j+s}) \varepsilon_{t-j}$$

which is identical to (8).

For the first order expectational difference equation, Muth's method works well. At the expense of some extra computation, it yields explicit expressions for C_j, though it is not clear that (14) is more useful than (7).

That Muth's method and that of section 1 are similar should not be surprising since they use the same logic but in two different spaces--Muth's in the space of sequences, the section 1 method in the space of analytic functions. But Muth's method becomes more complicated as the order of the equation increases, while the section 1 method does not. To see this, substitute (3) and (4) into (2) to get

$$\sum_{j=0}^{\infty} C_{j-1} \varepsilon_{t-j} - (\rho_1+\rho_2) \sum_{j=0}^{\infty} C_j \varepsilon_{t-j} + \rho_1\rho_2 \sum_{j=0}^{\infty} C_j \varepsilon_{t-j-1} = \sum_{j=0}^{\infty} A_j \varepsilon_{t-j}.$$

By matching up coefficients on ε_{t-j}, $j = 1, 2, \ldots$, one obtains

(16) $$C_{j+1} - (\rho_1+\rho_2)C_j + \rho_1\rho_2 C_{j-1} = (1-\rho_1 L)(1-\rho_2 L)C_j = A_{j-1} \qquad j=1,2,\ldots.$$

The equality for $j = 0$ is $C_1 - (\rho_1+\rho_2)C_0 = A_0$ which, along with $\sum_{j=0}^{\infty} C_j^2 < \infty$, can be considered a boundary condition for the second order difference equation (16). The general solution is

$$C_j = \{(1-\rho_1 L)(1-\rho_2 L)\}^{-1} A_{j-1} + k_1 \rho_1^j + k_2 \rho_2^j.$$

This expression is somewhat more complicated than the corresponding one for the first order case because of the presence of the product of polynomials in the lag operator. Sargent (1979b, Chapter IX) suggests applying the method of partial fractions to obtain[2]

$$C_j = \frac{\rho_1}{\rho_1-\rho_2} \frac{A_{j-1}}{1-\rho_1 L} - \frac{\rho_2}{\rho_1-\rho_2} \frac{A_{j-1}}{1-\rho_2 L} + k_1 \rho_1^j + k_2 \rho_2^j.$$

Suppose $|\rho_1| < 1$ and $|\rho_2| < 1$. Then using the advice given above,

$$C_j = \frac{\rho_1}{\rho_1-\rho_2} \sum_{s=0}^{\infty} \rho_1^s A_{j-1-s} - \frac{\rho_2}{\rho_1-\rho_2} \sum_{s=0}^{\infty} \rho_2^s A_{j-1-s} + k_1 \rho_1^j + k_2 \rho_2^j.$$

In this case, $\sum_{j=0}^{\infty} C_j^2 < \infty$ for any k_1 and k_2. Thus there is only one binding boundary condition, $C_1 - (\rho_1+\rho_2)C_0 = A_0$. Now

$$C_0 = k_1 + k_2$$

$$C_1 = \frac{\rho_1}{\rho_1-\rho_2} A_0 - \frac{\rho_2}{\rho_1-\rho_2} A_0 + k_1 \rho_1 + k_2 \rho_2$$

$$= A_0 + k_1 \rho_1 + k_2 \rho_2.$$

Then using the boundary condition,

$$C_1 = A_0 + k_1 \rho_1 + k_2 \rho_2 = A_0 + (\rho_1+\rho_2)C_0 = A_0 + (\rho_1+\rho_2)(k_1+k_2)$$

which gives $\rho_1 k_2 + \rho_2 k_1 = 0$. Thus k_1 and k_2 are not determined uniquely; each depends on C_0. The solution to (2) becomes

---

[2] It will be assumed that $\rho_1 \neq \rho_2$. The results below do not change in any essential way when $\rho_1 = \rho_2$.

$$y_t = \{ \sum_{j=0}^{\infty} \frac{\rho_1}{\rho_1-\rho_2} (\sum_{s=0}^{\infty} \rho_1^s A_{j-1-s} + C_0 \rho_1^j)$$

$$- \frac{\rho_2}{\rho_1-\rho_2} (\sum_{s=0}^{\infty} \rho_2^s A_{j-1-s} + C_0 \rho_2^j) \} \varepsilon_{t-j}$$

which is the analogue to (7).

When $|\rho_1| < 1$ and $|\rho_2| > 1$, the equation does have a unique solution. To see this, note that the appropriate expression for C_j is

$$C_j = \frac{\rho_1}{\rho_1-\rho_2} \sum_{s=0}^{\infty} \rho_1^s A_{j-1-s} + \frac{1}{\rho_1-\rho_2} \sum_{s=0}^{\infty} \rho_2^{-s} A_{j+s} + k_1 \rho_1^j + k_2 \rho_2^j.$$

Clearly, $\sum_{j=0}^{\infty} C_j^2 < \infty$ requires $k_2 = 0$. Then

$$C_0 = \frac{1}{\rho_1-\rho_2} A(\rho_2^{-1}) + k_1$$

$$C_1 = \frac{\rho_1}{\rho_1-\rho_2} A_0 + \frac{1}{\rho_1-\rho_2} \sum_{s=0}^{\infty} \rho_2^{-s} A_{1+s} + k_1 \rho_1$$

$$= \frac{\rho_1}{\rho_1-\rho_2} A_0 + \frac{1}{\rho_1-\rho_2} \rho_2[-A_0 + A_0 + \rho_2^{-1} A_1 + \rho_2^{-2} A_2 + \ldots] + k_1 \rho_1$$

$$= A_0 + \frac{\rho_2}{\rho_1-\rho_2} A(\rho_2^{-1}) + k_1 \rho_1.$$

Using $C_1 - (\rho_1+\rho_2)C_0 = A_0$,

$$k_1 = \frac{-\rho_1 \rho_2^{-1} A(\rho_2^{-1})}{\rho_1-\rho_2}$$

which implies $C_0 = -\rho_2^{-1} A(\rho_2^{-1})$ as in section 1. Then the solution to (2) can be written

$$y_t = \sum_{j=0}^{\infty} \{\frac{\rho_1}{\rho_1-\rho_2} (\sum_{s=0}^{\infty} \rho_1^s A_{j-1-s} - \rho_2^{-1} A(\rho_2^{-1})\rho_1^j) - \frac{\rho_2}{\rho_1-\rho_2} \sum_{s=0}^{\infty} \rho_2^{-s} A_{j+s}\} \varepsilon_{t-j}$$

which is the analogue to (8).

To see that there is no solution for $\{C_j\}$ satisfying $\sum_{j=0}^{\infty} C_j^2 < \infty$ when $|\rho_1| > 1$ and $|\rho_2| > 1$, notice that the appropriate expression for C_j in this case is

$$C_j = -\frac{1}{\rho_1-\rho_2} \sum_{s=0}^{\infty} \rho_1^{-s} A_{j+s} + \frac{1}{\rho_1-\rho_2} \sum_{j=0}^{\infty} \rho_2^{-s} A_{j+s} + k_1 \rho_1^j + k_2 \rho_2^j$$

while square-summability requires $k_1 = k_2 = 0$. But then

$$C_0 = \frac{1}{\rho_1-\rho_2} (-A(\rho_1^{-1}) + A(\rho_2^{-1}))$$

$$C_1 = A_0 - \frac{\rho_1 A(\rho_1^{-1})}{\rho_1-\rho_2} + \frac{\rho_2 A(\rho_2^{-1})}{\rho_1-\rho_2}.$$

With a little algebra, it can be shown that the condition $C_1 - (\rho_1+\rho_2)C_0 = A_0$ requires $\rho_2 A(\rho_1^{-1}) - \rho_1 A(\rho_2^{-1}) = 0$, a condition which will not hold in general.

The calculations above indicate that while Muth's technique can be used to obtain the solutions of section 1, it is somewhat more complicated than the one used there. In fact, Muth's technique can be characterized as a combination of the method of section 1 with an algorithm for inverting the z-transform. However, this combination is not an efficient one. While both techniques require the factorization of a polynomial, C_0 is determined immediately in section 1; Muth's technique requires an inversion operation (the solution of the difference equation in C_j) before C_0 can be determined using relatively complicated expressions involving convolutions. Since the inversion can be no more difficult when C_0 is known than when it is not, if one wants explicit expressions for $\{C_j\}$, the method of section 1 combined with a z-transform inversion technique may afford a considerable computational saving over Muth's. Of course, as was argued above, since much can be learned from C(z) itself, it is often not necessary to go this far.

3. Time Domain: Autoregressions

Lucas (1972a, 1975) has made use of a solution technique directed at calculating the sequence $\{d_j\}$ in (11) directly. Linear systems theorists will recognize this "method of undetermined coefficients in the time domain using autoregressions" as a "state-space" technique. To see why this terminology is appropriate, suppose $A(L) = (1-a_1 L)^{-1}$ so that the autoregressive representation for x_t is

(17) $\qquad x_t = a_1 x_{t-1} + \varepsilon_t .$

In the economy described by (1), the new information at time t is ε_t. The system "remembers" information which was new at time t-1, but via (17)--which agents are assumed to know. Thus the state of the economy at time t is x_t. Similarly, if

(18) $\qquad x_t = a_1 x_{t-1} + a_2 x_{t-2} + \varepsilon_t ,$

the state of the economy at time t is $\{x_t, x_{t-1}\}$.

Since (1) is linear, the solution y_t will be a linear function of the state at time t. This suggests that instead of (11), one look for solutions of the form

$$y_t = \sum_{j=0}^{N} d_j x_{t-j},$$

where N is one less than the order of the autoregression governing $\{x_t\}$. To illustrate the solution procedure, suppose the model is (1) and x_t is given by (17). Then the solution y_t will be of the form $y_t = d_0 x_t$. From (17), $E_t y_{t+1} = d_0 E_t x_{t+1} = d_0 a_1 x_t$. This can be used in (1) to get

$$d_0 a_1 x_t - \rho_1 d_0 x_t = x_t,$$

whereby $d_0 = (a_1 - \rho_1)^{-1}$. Thus the solution is

(19) $\qquad y_t = (a_1 - \rho_1)^{-1} x_t.$

In the calculations leading to (19), the size of ρ_1 did not matter. Indeed, when $|\rho_1| > 1$, (19) is identical to the unique solution (10) of section 1. But the calculations there indicated that if $|\rho_1| < 1$, the solution to (1) is

$$y_t = (1-\rho_1 L)^{-1}(L+C_0 A(L)^{-1})x_t.$$

In the first order autoregressive case this becomes

$$y_t = (1-\rho_1 L)^{-1}(L+C_0(1-a_1 L))x_t$$

where C_0 is any finite real number. If C_0 is chosen to be $(a_1-\rho_1)^{-1} = -\rho_1^{-1}$ xA(ρ_1^{-1}), (19) results. But there are clearly many other stationary solutions which Lucas's technique will not find.

Beyond concealing nonuniqueness, the technique has the unfortunate property of converting the expectational difference equation into a system of nonlinear equations. To see this, suppose x_t is governed by the second order autoregression (18). Then the solution will be of the form

$$y_t = d_0 x_t + d_1 x_{t-1}.$$

Accordingly,

$$E_t y_{t+1} = d_0 E_t x_{t+1} + d_1 x_t$$

$$= d_0 a_1 x_t + d_0 a_2 x_{t-1} + d_1 x_t.$$

Then (1) becomes

$$(d_0 a_1 + d_1) x_t + d_0 a_2 x_{t-1} - \rho_1 d_0 x_t - \rho_1 d_1 x_{t-1} = x_t.$$

Matching coefficients, one obtains

$$d_0 a_1 + d_1 - \rho_1 d_0 = 1$$

$$d_0 a_2 - \rho_1 d_1 = 0$$

which are two nonlinear equations in the two unknowns d_0 and d_1. In this case, it is straightforward but tedious to calculate

$$d_1 = (a_1 + a_2 \rho_1^{-1} - \rho_1)^{-1}$$

$$d_2 = a_2 \rho_1^{-1} (a_1 + a_2 \rho_1^{-1} - \rho_1)^{-1}.$$

But the size of the nonlinear system increases as the complexity of the $\{x_t\}$ process increases: when $\{x_t\}$ is characterized by an nth order autoregression, n nonlinear equations in n unknowns arise. Such systems can be quite difficult to solve, even if one knows that a solution exists. In addition, one must be willing to specify n at the outset, making general analyses of solutions impossible.

Lucas's procedure also becomes more complicated as the model does. Suppose the model is (2) and $\{x_t\}$ is governed by (17). A shrewd guess is that the state of the economy at t is $\{y_{t-1}, x_t\}$. Thus write $y_t = b_1 y_{t-1} + b_2 x_t$, noting that stationarity requires $|b_1| < 1$. Then

$$E_t y_{t+1} = b_1 y_t + b_2 a_1 x_t$$

$$= b_1^2 y_{t-1} + b_1 b_2 x_t + b_2 a_1 x_t.$$

Substituting for $E_t y_{t+1}$ and y_t, but not y_{t-1}, in (2), one obtains

$$b_1^2 y_{t-1} + (b_1 b_2 + b_2 a_1) x_t - (\rho_1 + \rho_2)(b_1 y_{t-1} + b_2 x_t) + \rho_1 \rho_2 y_{t-1} = x_t.$$

Equating coefficients on like objects,

$$b_1^2 - (\rho_1 + \rho_2) b_1 + \rho_1 \rho_2 = (b_1 - \rho_1)(b_1 - \rho_2) = 0$$

$$b_1 b_2 + b_2 a_1 - (\rho_1 + \rho_2) b_2 = 1.$$

Suppose $|\rho_1| < 1$ and $|\rho_2| < 1$. Then there are two solutions for b_1; $b_1 = \rho_1$ and $b_1 = \rho_2$. Using the former, for instance, $b_2 = (a_1 - \rho_1)^{-1}$. Thus the solution can be written as

(20a) $\qquad y_t = \rho_1 y_{t-1} + (a_1 - \rho_1)^{-1} x_t$

or

(20b) $\qquad y_t = \rho_2 y_{t-1} + (a_1 - \rho_2)^{-1} x_t.$

Evidently, the technique has found two possible solutions. But the entire class of solutions has not been found; one solution picks the C_0 of section 1 as $C_0 = -\rho_1^{-1} A(\rho_1^{-1})$, the other as $C_0 = -\rho_2^{-1} A(\rho_2^{-1})$. Of course, any finite C_0 yields a stationary solution. When $|\rho_1| < 1$ and $|\rho_2| > 1$, b_1 must be chosen as $b_1 = \rho_1$ to guarantee stationarity, and the solution is then given by (20a). But when $|\rho_1| > 1$ and $|\rho_2| > 1$, b_1 cannot be chosen to yield a stationary solution. Thus Lucas's technique readily discovers the existence problem in this second order case. Unfortunately, this result is not general. For suppose a term in y_{t-2} is appended to (2). Then under (17) the state of the economy at time t is $\{y_{t-1}, y_{t-2}, x_t\}$, and the solution will be of the form

$$y_t = b_1 y_{t-1} + b_2 y_{t-2} + b_3 x_t.$$

To find b_1 and b_2, two nonlinear equations must be solved. For stationarity, the roots of $1 - b_1 z - b_2 z^2$ must lie outside the unit circle. This condition will be quite difficult to check, as b_1 and b_2 will be complicated functions of the parameters of the expectational difference equation. In contrast, the stationarity condition in the method of section 1 involves only the raw parameters of the modified (2).

In addition to concealing existence and uniqueness problems, under certain conditions Lucas's technique is seemingly impossible to apply. Retaining (17), suppose (2) comes in the form

$$E_t y_{t+1} - \phi_1 y_t + \phi_2 y_{t-1} = x_t$$

and upon factoring the polynomial, $(1-\phi_1 z + \phi_2 z^2) = (1-\rho_1 z)(1-\rho_2 z)$, one finds ρ_1 and ρ_2 to be complex conjugates lying inside the unit circle. While the section 1 results indicate that a stationary solution exists, an attempt to find a real solution of the form

$$y_t = b_1 y_{t-1} + b_2 x_t$$

will fail, as b_1 will be complex. Though one might be tempted to give up at this stage, a solution of the form

(11) $$y_t = \sum_{j=0}^{\infty} d_j x_{t-j}$$

does exist. In fact, many such solutions exist, though $d_j \neq 0$ for all j in each. To find these solutions, substitute (11) into (2):

$$\sum_{j=0}^{\infty} d_{j+1} x_{t-j} - (\rho_1+\rho_2) \sum_{j=0}^{\infty} d_j x_{t-j} + \rho_1\rho_2 \sum_{j=0}^{\infty} d_j x_{t-1-j} = x_t.$$

Reference to the previous section indicates that Lucas's technique has become like Muth's, albeit somewhat simpler, for by matching coefficients one obtains

$$d_{j+1} - (\rho_1+\rho_2)d_j + \rho_1\rho_2 d_{j-1} = 0 \qquad j = 1,2,\ldots$$

with a single boundary condition given by $d_1 - (\rho_1+\rho_2)d_0 = 1$. Defining $r = |\rho_1|$ and $\theta = \tan^{-1}(\text{Re }\rho_1/\text{Im }\rho_1)$, the expression for d_j becomes

$$d_j = k_1 r^j \cos j\theta + k_2 r^j \sin j\theta \qquad j = 1,2,\ldots$$

where k_1 and k_2 depend (linearly) on d_0. Thus the lag weights display damped oscillations; y_t is given by

$$y_t = \sum_{j=0}^{\infty} (k_1(d_0) r^j \cos j\theta + k_2(d_0) r^j \sin j\theta) x_{t-j}.$$

The complications associated with (2) under Lucas's technique increase as the complexity of the $\{x_t\}$ process increases. For very simple cases, the technique is tractable. But for models of moderate size and interest, its sometimes bewildering complexity and issue-concealing properties seem to discourage its use when the method of section 1 is available.

4. Forward and Backward Solutions

Still another method of undetermined coefficients was suggested, for instance, by Blanchard (1978), and used by Fischer (1979). Like Lucas's technique, this method relies on a rather shrewd guess at the form of the solution. In (1), for instance, the guess is that y_t can be written in terms of past (backward) and expected future (forward) values of x:

(21) $$y_t = \sum_{j=1}^{\infty} g_j x_{t-j} + \sum_{j=0}^{\infty} h_j E_t x_{t+j}.$$

It is worthwhile to note that if $h_j \neq 0$ for $j > 0$, (21) is not an observable solution in the sense that $\sum_{j=0}^{\infty} h_j E_t x_{t+j}$ is an inexplicit function of the information available at time t. Thus, the technique converts the unobservable (1) into another unobservable, (21). Additional, sometimes considerable work is required to convert (21) into a function of $\{x_s\}_{s=-\infty}^{t}$.

The sequences $\{g_j\}$ and $\{h_j\}$ are obtained by using (21) to calculate $E_t y_{t+1}$,

$$E_t y_{t+1} = \sum_{j=1}^{\infty} g_j x_{t+1-j} + \sum_{j=0}^{\infty} h_j E_t x_{t+j+1}$$

and substituting this expression, along with (21), into (1):

$$\sum_{j=0}^{\infty} g_j x_{t+1-j} + \sum_{j=0}^{\infty} h_j E_t x_{t+j+1} - \rho_1 \sum_{j=1}^{\infty} g_j x_{t-j} - \rho_1 \sum_{j=0}^{\infty} h_j E_t x_{t+j} = x_t.$$

By equating coefficients on like elements of $\{x_t\}$, one obtains

$$g_1 - \rho_1 h_0 = 1$$

$$g_{j+1} - \rho_1 g_j = 0 \qquad j = 1, 2, \ldots$$

$$h_j - \rho_1 h_{j+1} = 0 \qquad j = 0, 1, 2, \ldots,$$

two difference equations connected by one boundary condition. Clearly, h_0 completely determines $\{g_j\}$ and $\{h_j\}$: the solutions are

$$h_j = h_0 \rho_1^{-j} \qquad j = 0,1,2,\ldots$$

$$g_j = \left(\frac{1+\rho_1 h_0}{\rho_1}\right)\rho_1^j \qquad j = 1,2,\ldots$$

so that (21) becomes

(22) $\qquad y_t = \left(\dfrac{1+\rho_1 h_0}{\rho_1}\right) \sum\limits_{j=1}^{\infty} \rho_1^j x_{t-j} + h_0 \sum\limits_{j=0}^{\infty} \rho_1^{-j} E_t x_{t+j}.$

Blanchard (1979) calls (22) with $h_0 = 0$ the "backward" solution; when $h_0 = -\rho_1^{-1}$, it is referred to as the "forward" solution.

Suppose $|\rho_1| < 1$. Then the forward solution will not exist in general and h_0 must equal zero. Then (22) becomes $y_t = \rho_1^{-1} \sum\limits_{j=1}^{\infty} \rho_1^j x_{t-j}$, which is equivalent to (7) with $C_0 = 0$. Since any finite value of C_0 determines a stationary solution, the forward-backward technique conceals the nonuniqueness problem when the forward solution does not exist. When it does exist, nonuniqueness is not concealed, but to obtain a closed form solution for y_t, the forward solution must be made observable. Virtually the only way to do this when $\{x_t\}$ is governed by the general moving average representation (3) is to use the techniques of section 1 to obtain the Hansen-Sargent (1980a) formula

$$\sum_{j=0}^{\infty} \rho_1^{-j} E_t x_{t+j} = \frac{I - \rho_1^{-1} L^{-1} A(\rho_1^{-1}) A(L)^{-1}}{1 - \rho_1^{-1} L^{-1}} x_t$$

where $A(\rho_1^{-1}) = \sum\limits_{j=0}^{\infty} A_j \rho_1^{-j}$ exists by assumption. Then (22) becomes

$$y_t = (1+\rho_1 h_0) \sum_{j=0}^{\infty} \rho_1^j x_{t-1-j} + h_0 \left[\frac{I - \rho_1^{-1} L^{-1} A(\rho_1^{-1}) A(L)^{-1}}{1 - \rho_1^{-1} L^{-1}} \right] x_t.$$

To find a closed form expression, assume $A(L)^{-1} = \sum_{j=0}^{q} \alpha_j L^j$. Then by another formula of Hansen and Sargent,

$$\frac{I - \rho_1^{-1} L^{-1} A(\rho_1^{-1}) A(L)^{-1}}{1 - \rho_1^{-1} L^{-1}} x_t = A(\rho_1^{-1}) x_t$$

$$+ A(\rho_1^{-1}) \sum_{j=1}^{q-1} \sum_{k=j+1}^{q} \rho_1^{-k+j} \alpha_k x_{t-j}.$$

Using this in (22) and collecting coefficients on like elements of $\{x_t\}$, one obtains a solution of the form

(11) $$y_t = \sum_{j=0}^{\infty} d_j x_{t-j}$$

where each d_j depends on h_0, ρ_1, and A_j. But the resulting $\{d_j\}$ sequence could have been obtained more simply by setting $C_0 = h_0 A(\rho_1^{-1})$ and inverting the z-transform implicit in (7).

When $|\rho_1| > 1$, the backward solution does not exist in the mean square sense. Although it might exist for a particular realization, one must ignore this possibility and set $h_0 = -\rho_1^{-1}$. Then (22) becomes

$$y_t = -\rho_1^{-1} \sum_{j=0}^{\infty} \rho_1^{-j} E_t x_{t+j}.$$

The first Hansen-Sargent formula converts this to

$$y_t = \frac{L - \rho_1^{-1} A(\rho_1^{-1}) A(L)^{-1}}{1 - \rho_1^{-1} L} x_t$$

which is the autoregressive version of (8). A closed form like (11) can be obtained by using the second Hansen-Sargent formula to get

$$y_t = A(\rho_1^{-1})x_t + A(\rho_1^{-1}) \sum_{j=1}^{q-1} \sum_{k=j+1}^{q} \rho_1^{-k+j} \alpha_k x_{t-j}.$$

Clearly, the forward-backward technique, like Muth's, combines the method of section 1 with an inversion algorithm. The advantage it has over Muth's is that it allows the investigator to make use of a known inversion, the Hansen-Sargent formula. The disadvantage it has relative to Muth's is that one must take great care with the forward solution to find evidence of nonuniqueness. But because the forward-backward technique is so like Muth's, it also becomes somewhat more complicated as the model does.

Applied to (2), the forward-backward technique finds only one solution regardless of ρ_1 and ρ_2. To see this, use (21) to calculate $E_t y_{t+1}$ and y_{t-1} and substitute the results into (2) to obtain

$$\sum_{j=0}^{\infty} g_{j+1} x_{t-j} + \sum_{j=0}^{\infty} h_j E_t x_{t+j+1} - (\rho_1+\rho_2)\sum_{j=1}^{\infty} g_j x_{t-j} + \sum_{j=0}^{\infty} h_j E_t x_{t+j}$$

$$+ \rho_1 \rho_2 \sum_{j=1}^{\infty} g_j x_{t-1-j} + \sum_{j=0}^{\infty} h_j E_{t-1} x_{t-1+j} = x_t.$$

Because of the last infinite series, $h_j = 0$ for all j. Then the equalities in g_j require

$$g_{j+1} - (\rho_1+\rho_2)g_j + \rho_1\rho_2 g_{j-1} = 0,$$

with boundary conditions $g_1 = 1$ and $g_2 - (\rho_1+\rho_2)g_1 = 0$. These two conditions uniquely determine g_j as

$$g_{j+1} = (\rho_2-\rho_1)^{-1}(\rho_2\rho_1^{j+1} - \rho_1\rho_2^{j+1}) \qquad j = 0,1,\ldots$$

whereby

$$y_t = (\rho_2-\rho_1)^{-1} \sum_{j=1}^{\infty} (\rho_2\rho_1^{j+1}-\rho_1\rho_2^{j+1})x_{t-j}.$$

This solution will be nonstationary unless $|\rho_1| < 1$ and $|\rho_2| < 1$, in which case it corresponds to the second-order version of (7) with $C_0 = 0$. Thus the technique provides no hint of the multitude of solutions when $|\rho_1| < 1$ and $|\rho_2| < 1$; more important, it fails to find the unique stationary solution under the configuration $|\rho_1| < 1 < |\rho_2|$.

It is a small consolation that the technique can be modified to handle (2). For suppose that the guess is written as

(23) $$y_t = by_{t-1} + \sum_{j=1}^{\infty} g_j x_{t-j} + \sum_{j=0}^{\infty} h_j E_t x_{t-j}$$

rather than as (21). As in the previous section, b is the appropriate root of $b^2 - (\rho_1+\rho_2)b + \rho_1\rho_2 = 0$. When there is one real root inside the unit circle, the use of (23) will lead to the unique solution. But the entire class of stationary solutions which exist when both roots are real and have less than unit modulus will only be found if the forward solution exists. To be fair, the use of (23) will immediately uncover the existence problem, but if the roots of the characteristic polynomial are complex, one is led into the same quagmire encountered in the previous section.

While the method of section 1 corresponds to one's intuition of an algorithm, the forward-backward technique, as well as that of Lucas, does not: an algorithm requires no shrewdness. But as has been seen, the successful use of either of the latter two techniques often requires a shrewd guess about the form of the solution. Even so, both techniques tend to obscure existence and uniqueness problems which the technique of section 1 finds readily.

5. Operator Techniques

Several authors, including Wallis (1980), Shiller (1978), and Sargent (1979b) have employed a technique which transforms the expectational difference equation into an ordinary difference equation, albeit one in the special sequences $E_t\{y_{t+j}\}_{j=1}^{\infty}$ and $E_t\{x_{t+j}\}_{j=1}^{\infty}$. Standard lag operator techniques are then used to solve for the expectations of interest. Equation (1) is appropriate for illustrating this technique, as it was used for expositional purposes by Wallis and Shiller.

The technique is implemented by leading (1) one period and taking conditional expectations as of time t to get

$$E_t(E_{t+1}y_{t+2} - \rho_1 y_{t+1}) = E_t x_{t+1}$$

or

$$E_t y_{t+2} - \rho_1 E_t y_{t+1} = E_t x_{t+1}$$

by the law of iterated expectations. Suppose $|\rho_1| > 1$ and define $\hat{y}_{t+j} = E_t y_{t+j}$ and $\hat{x}_{t+j} = E_t x_{t+j}$. The equation then becomes

(24) $$\hat{y}_{t+2} - \rho_1 \hat{y}_{t+1} = \hat{x}_{t+1}.$$

Both Shiller and Wallis solve this difference equation by forward substitution to get

(25) $$\hat{y}_{t+1} = -\rho_1^{-1} \sum_{j=0}^{\infty} \rho_1^{-j} \hat{x}_{t+j}.$$

This expression, made operational by the Hansen-Sargent formula, can be substituted into (1) to obtain the unique solution y_t. Relying on the "stability condition" $|\rho_1| > 1$, Wallis ignores the case $|\rho_1| < 1$. Shiller considers this case but remarks,[3] "The solution (25) is no longer applicable, since coefficients in the summation sum to infinity." The coefficients do sum to infinity, but, as reference to the previous section indicates, his conclusion is false. The solution he proposes for this case is one of the infinite number of such solutions; he proposes the backward solution--(7) with $C_0 = 0$.

As presented so far, this operator technique has two properties. First, one obtains an expression for $E_t y_{t+1}$, <u>not</u> y_t, which is obtained only after resubstitution into the expectational difference equation. Second, in the event that one obtains an interesting solution for $E_t y_{t+1}$, it involves an infinite series in expected future values of x--a series which must then be converted into an expression in current and past values of x. Yet the technique still seems a bit vague. The reason for this is that it is an attempt to use the logic of section 1 without using complex analysis. In other words, the "operator" technique is supposed to capture the essence of the section 1 method in the same way that lag operator techniques capture the essence of the series expansion method of solving ordinary difference equations. Yet while the rules and regulations for lag operators are well known, those for operators appropriate for expectational difference equations are not.

Both Shiller (1978) and Sargent (1979b) have attempted to define an appropriate operator, though Shiller's notation is markedly more complicated. Thus, following Sargent, define the operator B on sequences $\{E_t y_{t+j}\}$ by

$$BE_t y_{t+j} = E_t y_{t+j-1}, \quad B^{-1} E_t y_{t+j} = E_t y_{t+j+1}.$$

---

[3]Note 13, p. 31.

Using this operator, (1) becomes

$$(B^{-1}-\rho_1)E_t y_t = E_t x_t,$$

or

$$-\rho_1(1-\rho_1^{-1}B^{-1})E_t y_t = E_t x_t.$$

Provided $|\rho_1| > 1$, one simply inverts the operator $(1-\rho_1^{-1}B^{-1})$ in the same way one would invert $(1-\rho_1^{-1}L^{-1})$:

$$y_t = \frac{-\rho_1^{-1}}{1 - \rho_1^{-1}B^{-1}} E_t x_t = -\rho_1^{-1} \sum_{j=0}^{\infty} \rho_1^{-j} E_t x_{t+j}.$$

Unfortunately, the proviso was crucial. As Sargent (1979b, p. 337) notes,

> We must be careful here because the properties of B make the forward inverse of $[1 - \rho_1^{-1}B^{-1}]$ the only legitimate one, apart from reasons of convergence. Operating on both sides of an equation with polynomials in non-positive powers of B is legitimate. But it is <u>not</u> legitimate to operate with polynomials in positive powers of B. For example, $E_t x_{t+1} = E_t y_{t+1}$ does <u>not</u> imply that $BE_t x_{t+1} = BE_t y_{t+1}$, i.e., $x_t = y_t$.

But when $|\rho_1| < 1$, one would, for convergence reasons, write the inversion in positive powers of B. A complete description of the properties of the B operator thus requires knowledge of the method of section 1. But once that method is learned, it seems that there is little to be gained by using operators at all.

6. The Saracoglu-Sargent Technique

Saracoglu and Sargent (1978) used a frequency domain solution technique which relied on the covariance and cross-covariance generating functions of y and x rather than the z-transforms of the moving average coefficients. Their technique makes possible an elegant existence proof, though it somewhat complicates the search for the stationary solution itself. But the existence proof provides a striking illustration of the fixed point property of rational expectations solutions, while a comparison of a complete rendition of their method of undetermined coefficients with the method of section 1 demonstrates the simplicity of the latter.

The existence proof can be illustrated in the context of (1). The moving average representation of $\{x_t\}$ is given by (3) while the covariance generating function, under the assumption of unit innovation variance, is $g_x(z) = A(z)A(z^{-1})$. From Sargent (1979b, Chapter XI), the projection of y_{t+1} on the information set $\{x_s\}_{-\infty}^{t}$ available at t can be written

$$E_t y_{t+1} = \sum_{j=0}^{\infty} h_j x_{t-j} \equiv h(L)x_t,$$

where the z-transform of the sequence $\{h_j\}$ is given by

$$h(z) = \left[\frac{g_{yx}(z)z^{-1}}{g_x(z)}\right]_+$$

$$= \frac{1}{A(z)} \left[\frac{g_{yx}(z)z^{-1}}{A(z^{-1})}\right]_+$$

whenever $A(z)^{-1}$ is a polynomial in nonnegative powers of z; i.e., whenever $\{x_t\}$ has an autoregressive representation. To convert (1) into an expression involving this z-transform, use $E_t y_{t+1} = h(L)x_t$, multiply both sides by $x_{t-\tau}$, and take expectations to get

$$Eh(L)x_t x_{t-\tau} - \rho_1 Ey_t x_{t-\tau} = Ex_t x_{t-\tau} \qquad \tau = 0 \pm 1, \pm 2, \ldots.$$

Now multiply by z^τ and sum over τ to obtain

$$h(z)g_x(z) - \rho_1 g_{yx}(z) = g_x(z)$$

by the convolution property of the z-transform. Substituting the expression for h(z) from above, one obtains

$$\frac{1}{A(z)} \left[\frac{g_{yx}(z)z^{-1}}{A(z^{-1})} \right]_+ + g_x(z) - \rho_1 g_{yx}(z) = g_x(z)$$

or

$$g_{yx}(z) = \rho_1^{-1} \frac{1}{A(z)} \left[\frac{g_{yx}(z)z^{-1}}{A(z^{-1})} \right]_+ + g_x(z) - g_x(z)$$

$$= \rho_1^{-1} \frac{1}{A(z)} \left[\frac{g_{yx}(z)z^{-1}}{A(z^{-1})} \right]_+ + A(z)A(z^{-1}) - A(z)A(z^{-1})$$

$$= A(z^{-1})\rho_1^{-1} \left[\frac{g_{yx}(z)z^{-1}}{A(z^{-1})} \right]_+ - A(z)$$

giving

(26) $$\frac{g_{yx}(z)}{A(z^{-1})} = \rho_1^{-1} \left[\frac{g_{yx}(z)z^{-1}}{A(z^{-1})} \right]_+ - A(z).$$

But in the context of (1), $g_{yx}(z) = C(z)A(z^{-1})$, where $C(z)$ is the z-transform of the moving average coefficients for y_t. Thus the functional equation (26) can be written

$$C(z) = \rho_1^{-1}\{[C(z)z^{-1}]_+ - A(z)\}.$$

Of course, this is just another way of writing the expression of section 1:

$$\rho_1 C(z) = [C(z)z^{-1}]_+ - A(z)$$

$$= [C(z) - C_0]z^{-1} - A(z)$$

or

$$(1-\rho_1 z)C(z) = zA(z) + C_0,$$

which is (6). Therefore, a solution to the functional equation (26) is a solution, in the sense of section 1, to the expectational difference equation (1).

Consider the operator

(27) $\qquad T[C(z)] = \rho_1^{-1}\{[C(z)z^{-1}]_+ - A(z)\}$

defined on the space of functions $C(z)$ on the unit disk. Since $C(z)$ is a square-summable power series, the space in question is complete. In fact, it is a Hilbert space.[4] By a version of Parseval's relation, the norm of an element of this space can be written

$$\left(\sum_{j=0}^{\infty} c_j^2\right)^{\frac{1}{2}} = \left|\frac{1}{2\pi i}\oint |C(z)|^2 \frac{dz}{z}\right|^{\frac{1}{2}}$$

where $\oint$ denotes contour-integration about the unit circle. From (27), if $u(z)$ and $v(z)$ represent z-transforms of square-summable sequences,

$$T[u(z)] - T[v(z)] = \rho_1^{-1}\{[u(z)z^{-1}]_+ - A(z) - [v(z)z^{-1}]_+ + A(z)\}$$

$$= \rho_1^{-1}\{[u(z)z^{-1}]_+ - [v(z)z^{-1}]_+\}$$

$$= \rho_1^{-1}[(u(z) - v(z))z^{-1}]_+$$

since the "plussing" operator is linear. Now

$$[(u(z) - v(z))z^{-1}]_+ = (u_1-v_1) + (u_2-v_2)z + \ldots$$

so that

---

[4]See, for instance, de Branges and Rovnyak (1966), p. 8.

$$\|[(u(z) - v(z))z^{-1}]_+\| \le \|u(z) - v(z)\|.$$

Thus

$$\|T[u(z)] - T[v(z)]\| \le |\rho_1^{-1}| \, \|u(z) - v(z)\|.$$

Therefore T is a contraction mapping whenever $|\rho_1| > 1$. In this case, by the contraction mapping or Banach fixed point theorem, the functional equation (27), and thus (1), has a unique solution $C(z)$. Clearly, the contraction argument has the same force as the analyticity requirement used in section 1. Of course, when $|\rho_1| < 1$ the operator is an "expansion," and there may be many solutions to (27).

When the solution is unique, it can be represented, using (27), as

(28) $$C(z) = \rho_1^{-1}\{[C(z)z^{-1}]_+ - A(z)\}.$$

Saracoglu and Sargent used a method of undetermined coefficients to find $C(z)$. In the present context, their method runs as follows. Notice first that since $C(z)$ involves only nonnegative powers of z, $C(z)z^{-1}$ involves at most one term in negative powers of z. Thus write

$$C(z)z^{-1} = gz^{-1} + [C(z)z^{-1}]_+$$

where g (= C_0) is to be determined. Next, substitute this expression into (28) to get

$$C(z) = \rho_1^{-1} \{C(z)z^{-1} - gz^{-1} - A(z)\}$$

or

(29) $$\frac{C(z)}{A(z)} = -\rho_1^{-1}(1-\rho_1^{-1}z^{-1})^{-1} \{1 + \frac{gz^{-1}}{A(z)}\}.$$

Since x_t has been assumed to have an autoregressive representation, $A(z)^{-1}$, and thus $C(z)A(z)^{-1}$, involves only nonnegative powers of z. Evidently, g must be chosen to make the right-hand side of (29) one-sided in nonnegative powers of z. To do this, (29) must be written in a more convenient form. The details of this rewriting procedure, omitted by Saracoglu and Sargent, are at the heart of the method.

The expression in (29) which must be one-sided is

$$(1 - \rho_1^{-1} z^{-1})^{-1} \{1 + \frac{gz^{-1}}{A(z)}\} = \frac{1}{1-\rho_1^{-1} z^{-1}} + \frac{gz^{-1}}{(1-\rho_1^{-1} z^{-1})A(z)}$$

$$= 1 + \frac{\rho_1^{-1} z^{-1}}{1-\rho_1^{-1} z^{-1}} + \frac{g}{(1-\rho_1^{-1} z^{-1})zA(z)}.$$

Now $-\rho_1^{-1} z^{-1}(1-\rho_1^{-1} z^{-1})^{-1} = (1-\rho_1 z)^{-1}$, which involves only negative powers of z: the convergent expansion is

$$\frac{\rho_1^{-1} z^{-1}}{1-\rho_1^{-1} z^{-1}} = \rho_1^{-1} z^{-1} \sum_{j=0}^{\infty} \rho_1^{-j} z^{-j}.$$

Then $g(1-\rho_1^{-1} z^{-1})^{-1} z^{-1} A(z)^{-1}$ also involves negative powers of z. But this term can be written, using a partial fractions expansion, as

$$\frac{g}{(1-\rho_1^{-1} z^{-1})zA(z)} = \frac{gN_1}{(z-\rho_1^{-1})} + \frac{gN_2}{A(z)}.$$

Clearly, $gN_2 A(z)^{-1}$ involves only nonnegative powers of z. To determine N_1, multiply both sides of the above expression by $(z-\rho_1^{-1})$ and evaluate the result at ρ_1^{-1}:

$$\left.\frac{g(z-\rho_1^{-1})}{(1-\rho_1^{-1} z^{-1})zA(z)}\right|_{z=\rho_1^{-1}} = gN_1 + \left.\frac{gN_2(z-\rho_1^{-1})}{A(z)}\right|_{z=\rho_1^{-1}}$$

or $N_1 = A(\rho_1^{-1})^{-1}$, which exists since

$$A(z) = \sum_{j=0}^{\infty} A_j z^j$$

converges for all $|z| < 1$. Then (29) becomes

$$\frac{C(z)}{A(z)} = 1 + \frac{\rho_1^{-1} z^{-1}}{1-\rho_1^{-1} z^{-1}} + \frac{gA(\rho_1^{-1})}{z-\rho_1^{-1}} + \frac{gN_2}{A(z)}$$

$$= \frac{\rho_1^{-1} z^{-1} + gA(\rho_1^{-1})^{-1} z^{-1}}{1-\rho_1^{-1} z^{-1}} + 1 + \frac{gN_2}{A(z)}$$

which will be one-sided only if $gA(\rho_1^{-1})^{-1} + \rho_1^{-1} = 0$, or $g = C_0 = -\rho_1^{-1} A(\rho_1^{-1})$ as in section 1. The condition for $C(z)$ to be one-sided is the same as the condition for it to be analytic. In addition, the residue removal calculations of section 1 are transformed here into the evaluation of coefficients in a partial fractions expansion.

The preceding calculations show that the method of section 1 is simply a short cut to the Saracoglu-Sargent results: (6) is somewhat easier to deal with than (29); the residue calculations involve less algebra than partial fractions expansions. Of course, the two methods are equally adept at discovering the multiplicity of solutions which exist when $|\rho_1| < 1$: (29) is then one-sided for any g.

Conclusion

Each of the methods discussed in this Chapter lead to some sort of solution to an expectational difference equation. With little effort, the method of section 1 leads to a characterization of the solution. The expression thus obtained is not a closed form, though it is easily converted into one. Using a less efficient ordering of this process, Muth's technique leads directly to a reduced form. The forward-backward technique and that of Lucas both require shrewd guesses at the form of the solution and tend to obscure existence and uniqueness problems. In addition, each is computationally burdensome: Lucas's requires solution of a system of nonlinear algebraic equations while the forward-backward "solution" requires additional work to be useful. Finally, the operator technique is hit or miss--its use necessitates a good deal of insight into the form of the solution. Together, these ideas can be converted into what is at least a mildly convincing argument in favor of the method of section 1 when the simple computation of the solutions is taken to be a desideratum in its own right. But there is an additional argument: by making extensive use of the tools of time series analysis, the section 1 method is ideal for use in conjunction with a maximum likelihood procedure for estimating the parameters of the expectational difference equation under the cross-equation restrictions of rational expectations.

III

Confronting Nonuniqueness in Rational Expectations Models

Some rational expectations models possess more than one solution. This should come as no surprise, for most of the rational expectations models which have appeared to date are, in essence, difference equations without boundary conditions.[1] There are, however, different types of nonuniqueness. One stems from the self-fulfilling nature of rational expectations. Another arises only for certain values of the parameters of the model.

In two recent papers, Taylor (1977) and McCallum (1980) propose alternative methods for dealing with nonuniqueness. Taylor's method, the "minimum variance" condition, chooses the solution with the smallest variance, while McCallum's "minimum state variable" condition chooses the solution which depends on the fewest other variables. In applications, it will often be the case that these methods are either intractable or inapplicable.

In the analysis to follow, the two types of nonuniqueness are described in the context of a one-equation rational expectations model--an

---

[1] See, for instance, Black (1974), Brock (1975), and Taylor (1977).

expectational difference equation. Next, the two methods for resolving nonuniqueness are studied in some detail. Finally, an alternative method is proposed--one which amounts to doing nothing.

1. <u>A Simple Expectational Difference Equation: Parametric Nonuniqueness</u>

Using the notation of Chapter I, the simplest expectational difference equation is:

(1) $\quad E_t y_{t+1} - \rho y_t = x_t \qquad \rho \in R.$

The variable x_t is a representative element in the zero-mean linearly regular convariance stationary stochastic process (LRCSSP^0) $\{x_t\}$. This "driving process" x_t has a Wold representation given by:

(2) $\quad x_t = \sum_{j=0}^{\infty} A_j \varepsilon_{t-j}.$

Thus $\varepsilon_t = x_t - E[x_t, x_{t-1}, x_{t-2}, \ldots]$ has mean zero and is serially uncorrelated, $\sum_{j=0}^{\infty} A_j^2 < \infty$, and $A(z) = \sum_{j=0}^{\infty} A_j z^j$ is an analytic function on $|z| < 1$. $A(z)$ is assumed to have an inverse in nonnegative powers of z; that is, $\{x_t\}$ has an autoregressive representation. The variable y_t is a representative element in the process $\{y_t\}$ which is assumed to be a LRCSSP^0. The expectations term $E_t y_{t+1}$ is taken to mean the best linear forecast of y_{t+1} based on information available at time t, $\{y_t, y_{t-1}, \ldots\} \cup \{\varepsilon_t, \varepsilon_{t-1}, \ldots\}$.[2]

The restricted but interesting class of solutions lying in the space spanned by time independent square summable linear combinations of

---

[2] Since $\{x_t\}$ is assumed to have an autoregressive representation, the space spanned by $\{y_t, y_{t-1}, \ldots\} \cup \{\varepsilon_t, \varepsilon_{t-1}, \ldots\}$ is identical to that spanned by $\{y_t, y_{t-1}, \ldots\} \cup \{x_t, x_{t-1}, \ldots\}$.

$\{\varepsilon_t, \varepsilon_{t-1}, \ldots\}$ can be written as

(3) $\quad y_t = \sum_{j=0}^{\infty} C_j \varepsilon_{t-j} = C(L)\varepsilon_t,$

where $\sum_{j=0}^{\infty} C_j^2 < \infty$ and $C(z)$ is analytic on $|z| < 1$. To find these solutions, first evaluate the term $E_t y_{t+1}$ using (3) and the Wiener-Kolmogorov formula (see Chapter I):

$$E_t y_{t+1} = E_t[C(L)\varepsilon_{t+1}]$$

$$= \left[\frac{C(L)}{L}\right]_+ \varepsilon_t$$

$$= L^{-1}[C(L) - C_0]\varepsilon_t.$$

With this expression, (1) becomes

(4) $\quad L^{-1}[C(L) - C_0]\varepsilon_t - \rho C(L)\varepsilon_t = A(L)\varepsilon_t,$

which is assumed to hold for all realizations of $\{\varepsilon_t\}$. Thus, the sequences of coefficients in (4) must be identical; their z-transforms must be equal as analytic functions on the open unit disk. Thus

$$z^{-1}[C(z) - C_0] - \rho C(z) = A(z)$$

or

(5) $\quad C(z) = (1-\rho z)^{-1}\{zA(z) + C_0\}.$

When $|\rho| > 1$, $C(z)$ possesses an isolated singularity inside the unit circle at $z = \rho^{-1}$. But $C(z)$ can be made analytic on $|z| < 1$ by removing the singularity--by setting $C_0 = -\rho^{-1}A(\rho^{-1})$.[3] Thus when $|\rho| > 1$, the unique y_t

---

[3] See Chapter I for details.

in $H_{\bar{x}}(t)$ (the space spanned, as above, by $\{\varepsilon_t, \varepsilon_{t-1}, \ldots\}$) which solves (1) is

$$y_t = (1-\rho L)^{-1}\{LA(L) - \rho^{-1}A(\rho^{-1})\}\varepsilon_t.$$

When $|\rho| < 1$, (5) is an expression for a $C(z)$ which is analytic for any finite C_0. In this case, there are an infinite number of solutions y_t in $H_{\bar{x}}(t)$ which solve (1). These solutions are represented parametrically by

(6) $\qquad y_t = (1-\rho L)^{-1}\{LA(L) + C_0\}\varepsilon_t.$

Equation (6) exhibits "parametric nonuniqueness:" there is an infinity of solutions to (1) because when $|\rho| < 1$, the model is not sufficiently restrictive. However, the model (with $|\rho| < 1$) and the multiple solutions to it embodied in (6) provide a convenient basis for studying problems of uniqueness in rational expectations models.

2. The "Leading Indicator" Problem

Taylor (1977) produced an example of a rational expectations model with multiple solutions. His example was a version of (1) where y_t was the price level, ε_t was a serially uncorrelated composite error term, and $A(L) = 1$. Thus, multiple solutions to (1) corresponded to multiple equilibrium price distributions.

Taylor made two claims about the nonuniqueness problem. The first was that

> ... a widely publicized leading indicator of future prices, which is generated in a purely random fashion, can increase the variance of the equilibrium price distribution. If everyone expects that the indicator is leading, then it is impossible to statistically verify that the indicator is a will-o'-the-wisp. (pp. 1377-1378)

The second claim was that a "minimum variance" criterion could be applied to resolve both the parametric nonuniqueness and leading indicator problems. The point of this section is that the leading indicator problem is not a general consequence of nonuniqueness. The next section analyzes the minimum variance condition.

The leading indicator problem does not arise in the model (1), even though with $|\rho| < 1$ there are many (finite variance) solutions to it. Suppose agents believe that y_t evolves according to

(7) $\quad y_t = C(L)\varepsilon_t + d(L)v_t$

where $\{\varepsilon_t\}$ and $\{v_t\}$ are orthogonal, and $d(L) = \sum_{j=1}^{\infty} d_j L^j$ so that $\{v_t\}$ is a leading indicator. Note that

$$E_t y_{t+1} = E_t[C(L)\varepsilon_{t+1} + d_1 v_t + d_2 v_{t-1} + \ldots]$$

$$= L^{-1}[C(L) - C_0]\varepsilon_t + d_1 v_t + d_2 v_{t-1} + \ldots]$$

$$= L^{-1}[C(L) - C_0]\varepsilon_t + L^{-1} d(L) v_t.$$

Using this expression in (1), one obtains

$$L^{-1}[C(L) - C_0]\varepsilon_t + L^{-1} d(L) v_t - \rho C(L)\varepsilon_t - \rho d(L) v_t = A(L)\varepsilon_t.$$

Thus the z-transforms of the sequences in (7) must satisfy

$$(1-\rho z)C(z) = zA(z) + C_0$$

$$(1-\rho z)d(z) = 0.$$

Clearly, $d(z) = 0$, and y_t is given, as above, by (6). Thus, a model with multiple solutions need not admit a spurious leading indicator: such a situation is not consistent with equation (1).

A leading indicator does arise in slightly different models with multiple solutions. A model which mimics Taylor's is

(8) $\quad E_{t-1}y_{t+1} - \rho y_t = x_t.$

Equation (8) exhibits the "withholding" property (see Chapter I): more information (y_t) appears in the equation than in the information set (dated t-1) available to agents. To study the nonuniqueness problem, use (3) and (8) to obtain

$$L^{-1}[C(L) - C_0 - C_1 L]\varepsilon_t - \rho C(L)\varepsilon_t = A(L)\varepsilon_t$$

whereby

(9) $\quad C(z) = (1-\rho z)^{-1}\{zA(z) + C_0 + C_1 z\}.$

Apparently, there are two undetermined parameters in (9). But (9) is a functional equation of the form

(10) $\quad C_0 + C_1 z + \sum_{j=2}^{\infty} C_j z^j = (1-\rho z)^{-1}\{zA(z) + C_0 + C_1 z\}.$

Notice that $dC(z)/dz$ evaluated at $z = 0$ is C_1. Thus for consistency, the derivative of the right-hand side of (10), evaluated at $z = 0$, must equal C_1. Therefore

$$C_1 = \left. \frac{(1-\rho z)(zA'(z) + A(z) + C_1) - (zA(z) + C_0 + C_1 z)(-\rho)}{(1-\rho z)^2} \right|_{z=0}$$

$$= A(0) + C_1 + \rho C_0,$$

or $C_0 = -\rho^{-1}A(0)$. Thus (9) becomes

(11) $\quad C(z) = (1-\rho z)^{-1}\{zA(z) - \rho^{-1}A(0) + C_1 z\}$

and the class of solutions to (8) is given by

(12) $\quad y_t = (1-\rho L)^{-1}\{LA(L) - \rho^{-1}A(0) + C_1 L\}\varepsilon_t,$

where only one parameter, C_1, is undetermined.

The only difference between (1) and (8) is the dating of the information set used in the expectations term. The lagged information set in (8) results in the presence of the additional term $\rho^{-1}A(0)$ in (12). In fact, the additional term ensures the analyticity of the z-transform of the moving average coefficients of the LRCSSP^0 $\{E_{t-1}y_{t+1}\}$. The term appears because the characteristic equation associated with (8) has a root not just outside the unit circle at ρ^{-1}, but also one inside the unit circle, at $z = 0$. It is precisely this property which allows (8) to admit a spurious leading indicator. To see this, suppose agents believe that $\{v_t\}$ is a leading indicator for $\{y_t\}$, and thus that y_t is described by (7). Substituting (7) into (8), one obtains

$$L^{-1}[C(L) - C_0 - C_1]\varepsilon_t + L^{-1}[d(L) - d_1 L]v_t - \rho C(L)\varepsilon_t - \rho d(L)v_t = A(L)\varepsilon_t.$$

It is clear from this expression that $C(z)$ is given by (11) and that

$$(1-\rho z)d(z) = d_1.$$

Thus the class of solutions to (8) is given by

(13) $\quad y_t = (1-\rho L)^{-1}\{LA(L) - \rho^{-1}A(0) + C_1 L\}\varepsilon_t + (1-\rho L)^{-1}d_1 v_{t-1}$

where C_1 and d_1 are arbitrary. In (13), $\{v_t\}$ is a spurious leading indicator.

Both (8) and a version of (1) exhibit the nonuniqueness property. Yet (8) admits a spurious leading indicator while (1) does not. Model (1) does, however, admit a spurious "coincident" indicator. To see this, suppose that agents believe that y_t evolves according to

(14) $\quad y_t = C(L)\varepsilon_t + d(L)v_t$

where now $d(L) = \sum_{j=0}^{\infty} d_j L^j$, so that $\{v_t\}$ can be thought of as a coincident indicator. Substituting (14) into (1), for $|\rho| < 1$, and taking z-transforms, one discovers a $C(z)$ given by (5) and $(1-\rho z)d(z) = d_0$. Thus the class of solutions y_t is given by

$$y_t = (1-\rho L)^{-1}\{LA(L) + C_0\}\varepsilon_t + (1-\rho L)^{-1}d_0 v_t$$

where C_0 and d_0 are arbitrary.

The three indicator problems are suggestive in two ways. First, a model with multiple solutions need not admit a spurious leading indicator. Second, a spurious indicator is actually just a solution to the homogeneous equation. When initial values are specified for the sequence y_t, the indeterminacies disappear (see Whittle, 1963, p. 15.) When such boundary conditions are not present, some other side condition is necessary to eliminate these indeterminacies. One important side condition is analyzed in the next section.

3. The Minimum Variance Condition

Taylor's (1977) suggestion for dealing with a multiplicity of solutions was to pick the one with the smallest variance. As will be seen below, the application of this minimum variance condition has a number of interesting implications.

To illustrate Taylor's technique, return to the model (8) and its solutions given by (13):

$$y_t = (1-\rho L)^{-1}\{LA(L) - \rho^{-1}A(0) + C_1 L\}\varepsilon_t + (1-\rho L)^{-1}d_1 v_{t-1}.$$

In Taylor's example, the driving process was itself white noise, so that $A(L) \equiv I$. Thus

$$y_t = (1-\rho L)^{-1}\{\rho^{-1}(1-\rho L) + C_1 L\}\varepsilon_t + (1-\rho L)^{-1} d_1 v_{t-1}$$

$$= -\rho^{-1}\varepsilon_t + C_1 \sum_{j=1}^{\infty} \rho^j \varepsilon_{t-j} + d_1 \sum_{j=1}^{\infty} \rho^{j+1} v_{t-j}.$$

The variance of y_t is then given by

$$Ey_t^2 = \sigma_\varepsilon^2(\rho^{-2} + C_1^2 \sum_{j=1}^{\infty} \rho^{2j}) + \sigma_v^2 d_1^2 \sum_{j=1}^{\infty} \rho^{2(j+1)},$$

which is smallest when $C_1 = d_1 = 0$. Taylor made two observations about this example. First, the minimum variance conditions on the two unknown parameters C_1 and d_1 are the same. Second, the condition on C_1 is the same as it would have been had (12) been unique: using (12) with $A(L) \equiv I$, $C(z)$ is

$$C(z) = (1-\rho z)^{-1}\{z - \rho^{-1} + C_1 z\},$$

which is not unique unless $|\rho| > 1$. In this case, C_1 must be set so as to make $C(z)$ analytic at ρ^{-1}:

$$\lim_{z \to \rho^{-1}} (1-\rho z)C(z) = 0 = C_1 \rho^{-1}$$

which requires $C_1 = 0$ as above. Neither of these observations is general, as analysis of the spurious indicator solution to model (1) will indicate.

The solution to (1) with $|\rho| < 1$ and $A(L) \equiv I$ is

$$y_t = (1-\rho L)^{-1}\{L + C_0\}\varepsilon_t + (1-\rho L)^{-1} d_0 v_t.$$

Because $\{\varepsilon_t\}$ and $\{v_t\}$ are orthogonal, minimum variance clearly requires $d_0 = 0$. Once d_0 is set to zero, the determination of C_0 requires knowledge of the variance of y_t. To find this variance, first calculate $(1-\rho L)^{-1}(C_0+L)$:

$$(1-\rho L)^{-1}(C_0+L) = C_0 \sum_{j=0}^{\infty} \rho^j L^j + L \sum_{j=0}^{\infty} \rho^j L^j$$

$$= C_0 \sum_{j=0}^{\infty} \rho^j L^j + \sum_{j=1}^{\infty} \rho^{j-1} L^j$$

$$= C_0 + (C_0\rho+1) \sum_{j=0}^{\infty} \rho^j L^{j+1}.$$

Thus y_t is given by

$$y_t = C_0 \varepsilon_t + (C_0\rho+1)\varepsilon_{t-1} + (C_0\rho+1)\rho\varepsilon_{t-2} + \ldots$$

and its variance is

$$Ey_t^2 = \sigma_\varepsilon^2 (C_0^2 + (C_0\rho+1)^2 + (C_0\rho+1)^2\rho^2 + \ldots)$$

$$= \sigma_\varepsilon^2 (C_0^2 + (C_0\rho+1)^2 \sum_{j=0}^{\infty} \rho^{2j})$$

$$= \sigma_\varepsilon^2 (C_0^2 + \frac{(C_0\rho+1)^2}{1-\rho^2}).$$

This quantity is minimized when its derivative with respect to C_0 is equal to zero:

$$\frac{dEy_t^2}{dC_0} = 2C_0 + \frac{2\rho(C_0\rho+1)}{1-\rho^2} = 0,$$

which requires $C_0 = -\rho$. Clearly, the variance-minimizing choices of C_0 and d_0 are not the same. In addition, the variance-minimizing choice of C_0 is not the same as the one which removes the singularity (outside the unit circle) at ρ^{-1} in $C(z)$. Such a procedure sets C_0 according to

$$\lim_{z \to \rho^{-1}} \{z + C_0\} = 0$$

or $C_0 = -\rho^{-1}$. Thus one cannot confront a model with a single procedure (a strategy for setting C_0) and hope to find the unique solution when it exists and the minimum variance solution when there are many finite variance solutions.

When the driving process is not white noise, the minimum variance condition can be applied using a technique developed by Nerlove, Grether, and Carvalho (1979, Chapter IV.3). To illustrate this technique, suppose the model is (1) with $|\rho| < 1$ and $A(L) = \sum_{j=0}^{\infty} A_j L^j$. Then the solution y_t can be written $y_t = (1-\rho L)^{-1}(C_0 + LA(L))\varepsilon_t$. Define a new process w_t by $w_t = (1-\rho L)^{-1}\varepsilon_t$. Also, let $a_0 = C_0$, $a_1 = A_0$, $a_2 = A_1, \ldots$, so that $C_0 + LA(L) = \sum_{j=0}^{\infty} a_j L^j \equiv a(L)$. Then $y_t = a(L)w_t$. The covariance generating function $g_y(z)$ of y_t is (see Sargent, 1979b, Chapter XI)

$$g_y(z) = a(z)a(z^{-1})g_w(z).$$

The covariance $Ey_t y_{t-\tau} \equiv C_y(\tau)$ is given by the coefficient on z^τ in $g_y(z)$. The simple process w_t has $g_w(z) = (1-\rho^2) \sum_{j=-\infty}^{\infty} \rho^{|\tau|} z^\tau$. Thus the variance of y_t is given by

$$Ey_t^2 = \frac{\rho^0}{1-\rho^2} \sum_{j=0}^{\infty} a_j^2 + \frac{2\rho}{1-\rho^2} \sum_{j=0}^{\infty} a_j a_{j+1} + \ldots$$

$$+ \frac{2\rho^k}{1-\rho^2} \sum_{j=0}^{\infty} a_j a_{j+k} + \ldots .$$

This quantity is minimized when its derivative with respect to a_0 (C_0) is equal to zero:

$$\frac{dEy_t^2}{da_0} = \frac{2a_0}{1-\rho^2} + \frac{2\rho}{1-\rho^2} \sum_{j=0}^{\infty} a_{j+1} \rho^j$$

$$= \frac{2C_0}{1-\rho^2} + \frac{2\rho}{1-\rho^2} \sum_{j=0}^{\infty} \rho^j A_j = 0$$

where the infinite series converges by virtue of the fact that $A(z)$ is analytic on $|z| < 1$. Thus, the variance-minimizing choice of C_0 is $-\rho A(\rho)$. When C_0 is set in this fashion, the solution y_t is given by $y_t = (1-\rho L)^{-1} \{LA(L) - \rho A(\rho)\} \varepsilon_t$. This minimum variance solution for $|\rho| < 1$ is clearly similar to the unique solution $y_t = (1-\rho L)^{-1}$ $\times \{LA(L) - \rho^{-1} A(\rho^{-1})\} \varepsilon_t$ appropriate when $|\rho| > 1$. The minimum variance solution does, however, have one additional property: it fails to have an autoregressive representation. The reason for this is that $C(z)$ has a zero inside the unit circle at $z = \rho$: $C(\rho) = (1-\rho^2)^{-1} \{\rho A(\rho) - \rho A(\rho)\} = 0$. Thus, $C(z)$ does not have an inverse in nonnegative powers of z, and y_t fails to have an autoregressive representation. This need not be true for the unique solution when $|\rho| > 1$. Indeed, when $|\rho| > 1$,

$$\lim_{z \to \rho^{-1}} C(z) = \lim_{z \to \rho^{-1}} \frac{zA(z) - \rho^{-1} A(\rho^{-1})}{1 - \rho z}$$

$$= \lim_{z \to \rho^{-1}} \{A(z) + zA'(z)\}/(-\rho)$$

$$= \{A(\rho^{-1}) + \rho^{-1}A'(\rho^{-1})\}/(-\rho)$$

by L'Hopital's rule. The terms on the right can be rearranged to give $C(\rho^{-1}) = \{2A(\rho^{-1}) - A_0\}/(-\rho)$, which will be zero only in very special cases.[4]

Besides producing noninvertible solutions, the minimum variance condition has yet another drawback: it proves useful only when there is a single undetermined parameter in $C(z)$. The reason is that the minimum variance criterion provides only one nonlinear equation in the unknown parameters. It would not, for instance, be sufficient to determine a solution to the model

$$E_t y_{t+2} - (\rho_1 + \rho_2) E_t y_{t+1} + \rho_1 \rho_2 y_t = x_t$$

where $|\rho_1|$, $|\rho_2| < 1$. In this example, $C(z)$ is given by

$$C(z) = [(1-\rho_1 z)(1-\rho_2 z)]^{-1}\{z^2 A(z) + C_0 + (C_1 + (\rho_1+\rho_2)C_0)z\}$$

where neither C_0 nor C_1 is determined. One could add a "smoothness" condition that the first covariance be maximized given a minimum variance, but this procedure is not likely to be useful in the presence of higher order nonuniqueness.

Although the minumum variance criterion has enjoyed some popularity, it has its drawbacks.[5] As has been seen, it may produce a noninvertible solution and is useful only for "first order" nonuniqueness. In addition, it does not imply that the parameters of the spurious indicator be set in the

---

[4] See Futia (1981) for a very interesting special case.

[5] Shiller (1978) and Gourieroux, Laffont, and Monfort (1982) discuss the condition in some detail.

same way as the parameters of the initial solution. Finally, it does not cause the solution that would have resulted had the parameters of the model forced a unique solution.

4. The Minimum State Variable Technique

McCallum (1980) has proposed a solution technique which inherently eliminates the nonuniqueness and spurious indicator problems. His method, like that of Lucas (1972a), is a method of undetermined coefficients using a minimal set of state variables. To illustrate this technique, suppose the model is (1) with $|\rho| < 1$ and $A(L) = (1-\lambda L)^{-1}$ so that x_t is a first order autoregression: $x_t = \lambda x_{t-1} + \varepsilon_t$. Lucas and McCallum conjecture a solution of the form $y_t = \pi x_t$, where x_t is thought of as a state variable. Using the guess, one obtains

$$E_t y_{t-1} - \rho y_t = \pi \lambda x_t - \rho \pi x_t$$

which must equal x_t. Thus $\pi(\lambda-\rho) = 1$ and $\pi = (\lambda-\rho)^{-1}$. Thus the solution y_t is given by

$$y_t = (\lambda-\rho)^{-1} x_t$$
$$= (\lambda-\rho)^{-1} (1-\lambda L)^{-1} \varepsilon_t.$$

The solution technique clearly eliminates the spurious indicator problem: a spurious indicator cannot be an element of the smallest set of variables from which a solution can be calculated. Further, it eliminates the parametric nonuniqueness discussed above. Indeed, of the infinity of solutions of the form

$$y_t = (1-\rho L)^{-1} \{L(1-\lambda L)^{-1} + C_0\} \varepsilon_t,$$

the minimum state variable technique chooses the one corresponding to $C_0 = (\lambda-\rho)^{-1}$.

Several points should be made about this technique. First, the minimum state variable condition is not the same as the minimum variance condition. This is easily seen for the simple case $A(L) = 1$. From above, the moving average representation for y_t in this case is

$$y_t = C_0\varepsilon_t + (C_0\rho+1)\varepsilon_{t-1} + (C_0\rho+1)\rho\varepsilon_{t-2} + \dots .$$

The minimum state variable is $x_t = \varepsilon_t$. Thus McCallum chooses C_0 to minimize the number of <u>terms</u> on the right-hand side of the above expression, or, equivalently, to minimize the number of <u>terms</u> in the expression for the variance of y_t:

$$Ey_t^2 = \sigma_\varepsilon^2(C_0^2 + (C_0\rho+1)^2 + (C_0\rho+1)^2\rho^2 + \dots).$$

The minimum state variable condition requires $C_0\rho + 1 = 0$, or $C_0 = -\rho^{-1}$. But from above, the variance minimizing choice of C_0 is $C_0 = -\rho$.

The second point about this technique was made by Taylor but bears repeating: "... it is possible to obtain a solution but not recognize it as one of many solutions, the others being obscured in the functional forms that are ruled out a priori" (Taylor 1977, note 7.) For instance, in the calculations above all solutions of the form $(1-\rho L)^{-1}\{LA(1-\lambda L)^{-1} + C_0\}\varepsilon_t$ other than $\pi(1-\lambda L)^{-1}$ are ruled out.[6] On the other hand, the z-transform technique clearly displays the multiplicity of solutions.

The third point concerns the computational burden of the minimum state variable condition. As the driving process becomes more complex, the number of (minimal) state variables increases. For instance, when the driving

---

[6]Lucas (1972a) notes the restrictive nature of the technique in note 10.

process is an nth order autoregression, the minimal state variables are x_t, $x_{t-1},\ldots,x_{t-(n-1)}$, and the solution is of the form

$$y_t = \sum_{j=0}^{n-1} \pi_j x_{t-j}.$$

The model imposes n nonlinear restrictions on the π_j^s. Similarly, when x_t is an nth order moving average, the minimal state variables are ε_t, $\varepsilon_{t-1},\ldots,\varepsilon_{t-(n-1)}$ and the solution is of the form

$$y_t = \sum_{j=0}^{n-1} \pi_j \varepsilon_{t-j}.$$

Again, a system of n nonlinear equations in $\pi_0,\ldots,\pi_{n-1}$ arises. In practice, such nonlinear systems are quite difficult to solve. This computational burden can be avoided by using the z-transform technique, which does not become more complicated as the complexity of the driving process increases.

5. A Restrictive Solution Concept

Neither of the uniqueness-resolving conditions appears to dominate the other. One, the minimum variance condition, is generally inapplicable. The other, the minimum state variable condition, is generally intractable. If these problems are to be avoided, either of two additional strategies might be adopted. The first is to require that the solution hold for a given value of y_t, say at $t = t_0$. This eliminates the spurious indicator problem as well as most parameter-induced nonuniqueness. This procedure, of course, amounts to using an initial condition to determine a solution. It is rare, however, for an ad hoc rational expectations model like (1) to come equipped with such an initial condition.

The second strategy is to require that solutions be functions of the objective features of the environment. That is, to require solutions y_t to lie in $H_{\bar{x}}(t)$.[7] This strategy eliminates spurious indicators, but not parameter-induced nonuniqueness. The resulting solutions have two nice properties.

The first property of such solutions is that their existence and uniqueness properties are completely determined by the parameters of the model. Thus, for a particular setting of parameters, a model may have one solution or equilibrium, many solutions, or no solution. But this property may well be a desirable one. Wallace (1980) has argued, for instance, that in a good model of fiat money, monetary equilibria are tenuous: such equilibria simply do not exist for certain settings of parameters. Wallace goes on to argue that multiple equilibria need not be a cause for concern:

> By requiring that beliefs be tied to objective features of the environment, rational expectations equilibria allow us to focus on a subset of the equilibria consistent with all unexplained beliefs. The fact that this subset turns out to contain more than one equilibrium does not provide an argument for abandoning the rational expectations equilibrium concept. Instead, it suggests that we look for a principle that justifies focusing on a subset of the rational expectations equilbria. We know already that we will not be able to justify focusing on equilibria not in the subset of rational expectations equilibria. (p. 55)

The minimum variance and minimum state variable conditions are principles, albeit possibly unsatisfactory ones, which justify focusing on a subset of rational expectations equilibria. The requirement that solutions lie in $H_{\bar{x}}(t)$ is also such a principle, though it does not always justify focusing on a single solution. But all admissible solutions have a second desirable property.

The requirement that solutions to models like (1) lie in the space spanned by the driving process is restrictive econometrically. For instance,

---

[7]Taylor (1977, p. 1378) notes that "if everyone chooses to ignore the [spurious] indicator (which is the rational thing to do when everyone expects everyone else to ignore it), then the indicator will not increase the variance of the price level."

suppose an econometrician wishes to estimate the single parameter in the model (1) and to test the cross-equation restrictions implied by rational expectations. The first step is to estimate an unrestricted moving average representation for (x_t, y_t) of the form

$$\begin{bmatrix} x_t \\ y_t \end{bmatrix} = \begin{bmatrix} A(L) & B(L) \\ C(L) & D(L) \end{bmatrix} \begin{bmatrix} \varepsilon_{1t} \\ \varepsilon_{2t} \end{bmatrix}$$

where $A(L)$, $B(L)$, $C(L)$, and $D(L)$ have a, b, c, and d parameters. When the solution y_t is not unique ($|\rho| < 1$), the restricted moving average representation implied by (1) is (see Chapter I):

$$\begin{bmatrix} x_t \\ y_t \end{bmatrix} = \begin{bmatrix} A(L) & B(L) \\ \dfrac{LA(L) + C_0}{1 - \rho L} & \dfrac{LB(L) + D_0}{1 - \rho L} \end{bmatrix} \begin{bmatrix} \varepsilon_{1t} \\ \varepsilon_{2t} \end{bmatrix}$$

Though the unrestricted system has a + b + c + d free parameters, the restricted system has only a + b + 3. Thus, so long as c + d > 3, the restricted system is overidentified; ρ may be estimated and the cross-equation restrictions tested by conventional maximum likelihood techniques.[8]

Conclusion

Rational expectations models may fail to have unique solutions for a variety of reasons. Parametric nonuniqueness occurs when the parameters of the model admit more than one solution. Another type of nonuniqueness

---

[8] When the solution is unique, the system is not much more identified: the condition is c + d > 1.

results if agents believe that a nominally extraneous variable helps forecast the future. Thus, solutions to rational expectations models may depend on any number of these "spurious indicators." One way to confront such nonuniqueness is to choose the solution with the smallest variance. Another is to choose the solution which depends on the fewest state variables. But these methods were shown to be, in general, either inapplicable or intractable. An alternative is to ignore spurious indicators while admitting parametric nonuniqueness. This alternative strategy produces multiple, econometrically tractable solutions whenever the environment, embodied in the parameters of the model, so dictates.

IV

Moving Average Representations for Multivariate Rational Expectations Models

Methods for solving multivariate rational expectations models have recently been studied by, among others, Blanchard and Kahn (1980), Chow (1981), Fair and Taylor (1980), and Futia (1981). But their procedures do not generally provide the solution in a form convenient for, say, estimation. The analysis herein exploits the properties of polynomial matrices to establish conditions for the existence and uniqueness of solutions to multivariate linear rational expectations models. But more important, a method similar to the one employed in Chapter I is utilized to find "closed form" expressions for solutions when they exist. Furthermore, the method used here is suggestive of ways to deal with multivariate "withholding" equations.

1. An Existence and Uniqueness Theorem

The model to be studied is

(1) $$E_t(\sum_{j=0}^{n} F_j L^{-j} + \sum_{j=1}^{m} G_j L^j) y_t = x_t$$

where L is the lag operator: $L^m y_t = y_{t-m}$, y_t and x_t are (qx1), F_j and G_j are (qxq), and x_t is a covariance stationary vector stochastic process with Wold moving average representation

(2) $\quad x_t = \sum_{j=0}^{\infty} A_j \varepsilon_{t-j} \equiv A(L)\varepsilon_t.$

Thus $\varepsilon_t = x_t - E[x_t | x_{t-1}, x_{t-2}, \ldots]$, and each element of $\sum_{j=0}^{\infty} A_j A_j'$ exists. Hansen and Sargent (1981a) have studied the case where n = m and $F_j = \beta^j G_j$ for all $j \neq 0$. In this case, (1) can be thought of as an Euler equation arising from a multiple-variable, linear-quadratic, stochastic, dynamic optimization problem. But in this more general form, (1) need not be such a first-order condition; it is more appropriately thought of as a model postulated, say, at the level of supply and demand equations.

For the present purpose, a solution y_t to (1) is of the form

(3) $\quad y_t = \sum_{j=0}^{\infty} C_j \varepsilon_{t-j} \equiv C(L)\varepsilon_t$

where $\{y_t\}$ can be taken to be stationary; i.e., the function C(L) is one-sided in nonnegative powers of L. This solution must be appropriate for all realizations of the vector innovation process $\{\varepsilon_t\}$. The conditions for the existence and uniqueness of such solutions are given in the following theorem.

Theorem. Suppose the model is (1) and that the moving average representation for x_t is given by (2). Suppose further that F_n is of full rank, that the roots of $\det[z^n(\sum_{j=0}^{n} F_j z^{-j} + \sum_{j=1}^{m} G_j z^j)] = \sum_{j=0}^{p} f_j z^j$ are distinct, and that rq of these roots are inside the unit circle while the other $p - rq \leq (n+m)q - rq$ roots lie outside the unit circle. Then

(i) if $r < n$, there are many solutions to (1) of the form (3)

(ii) if $r = n$, there is one solution to (1) of the form (3)

(iii) if $r > n$, there is no solution to (1) of the form (3).

Proof: To begin, write (1) as

$$E_t(F_0 y_t + F_1 y_{t+1} + \ldots + F_n y_{t+n} + G(L) y_t) = x_t$$

where $G(L) = \sum_{j=1}^{m} G_j L^j$. From (3) and the Wiener-Kolmogorov formula,

$$E_t y_{t+j} = \left[\frac{C(L)}{L^j}\right]_+ \varepsilon_t$$

$$= L^{-j}(C(L) - \sum_{i=0}^{j-1} C_i L^i) \varepsilon_t.$$

Substituting this expression and (2) into (1), one obtains

$$F_0 C(L) \varepsilon_t + F_1 L^{-1}(C(L) - C_0) \varepsilon_t + \ldots + F_n L^{-n}(C(L) - \sum_{j=0}^{n-1} C_j L^{-j}) \varepsilon_t + G(L) \varepsilon_t = A(L) \varepsilon_t$$

which must hold for all realizations of $\{\varepsilon_t\}$. Thus the coefficient matrices are related by the z-transform identities

(4)
$$(F(z^{-1}) + G(z))C(z) = A(z) + F_1 C_0 z^{-1} + \ldots + F_n \sum_{j=0}^{n-1} C_j z^{-n-j}$$

$$= A(z) + \sum_{i=1}^{n} \sum_{s=i}^{n} F_s C_{s-i} z^{-i}$$

$$= A(z) + \sum_{j=0}^{n-1} (\prod_{s=j+1}^{n} F_s z^{-s}) C_j.$$

By assumption, the determinant of the polynomial matrix $z^n(F(z^{-1}) + G(z))$ has no roots on the unit circle. Then by the corollary to Theorem 1 of the Appendix, the matrix can be factored as

$$z^n(F(z^{-1}) + G(z)) = S(z)T(z)$$

where $S(z)$ is a polynomial matrix such that all the roots of det $S(z)$ lie inside the unit circle while $T(z)$ is a polynomial matrix with all its roots outside the unit circle. $S(z)$ is nonsingular almost everywhere in the complex plane, and is of order r:

$$S(z) = \sum_{j=0}^{r} S_j z^r.$$

Let $k = rq$. Then det $S(z) = \sum_{j=0}^{k} d_j z^j = d_k \prod_{j=1}^{k} (z-z_j)$ where the roots z_j lie inside the unit circle and are distinct. The coefficients d_j can be found using the algorithm described by Hansen and Sargent (1981a). Now

$$S(z)^{-1} = \frac{\text{Adj } S(z)}{\det S(z)}$$

where Adj $S(z)$ is of order k-r:

$$\text{Adj } S(z) = \sum_{j=0}^{k-r} S_j^* z^j.$$

The partial fractions expansion of $S(z)^{-1}$ can then be written

$$S(z)^{-1} = \sum_{j=1}^{k} \frac{N_j}{z-z_j}$$

where

$$N_j = \frac{\text{Adj } S(z_j)}{d_k \prod_{\substack{i \neq j \\ 1 \leq i \leq k}} (z_i - z_j)} \qquad j = 1, \ldots, k.$$

By the corollary to Theorem 2 of the Appendix, rank $(N_j) = 1$ for $j = 1, \ldots, k$. Using the expansion of $S(z)^{-1}$, the z-transform identities (4) become

(5) $$T(z)C(z) = \sum_{j=1}^{k} \frac{N_j}{z - z_j} \{z^n A(z) + z^n \sum_{i=1}^{n} \sum_{s=i}^{n} F_s C_{s-i} z^{-i}\}$$

which is valid for all z on the open unit disk except $z = z_j$, $j = 1, \ldots, k$. But since $C(z)$ is the z-transform of the moving average coefficients for y_t, it must exist for all $|z| < 1$. This condition places restrictions on the nq^2 unknown parameters $C_0, C_1, \ldots, C_{n-1}$. Indeed, because the right-hand side of (5), as it stands, is undefined for $z = z_j$, $j = 1, \ldots, k$, it must be the case that

$$(z - z_j)T(z)C(z)\big|_{z=z_j} = 0,$$

or

(6) $$N_j\{z_j^n A(z_j) + z_j^n \sum_{i=1}^{n} \sum_{s=1}^{n} F_s C_{s-i} z_j^{-i}\} = 0 \qquad j = 1, \ldots, k.$$

Using (4), expression (6) can be written as

(7) $$\begin{bmatrix} N_1 z_1^n A(z_1) \\ \vdots \\ N_k z_k^n A(z_k) \end{bmatrix} = - \begin{bmatrix} N_1 z_1^n \sum_{i=1}^{n} F_i z_1^{-1} & \cdots & N_1 z_1^n F_n z_1^{-n+1} \\ \vdots & & \vdots \\ N_k z_k^n \sum_{i=1}^{n} F_i z_k^{-1} & \cdots & N_k z_k^n F_n z_k^{-n+1} \end{bmatrix} \begin{bmatrix} C_0 \\ \vdots \\ C_{n-1} \end{bmatrix}$$

$\qquad\qquad$ (kq × q) $\qquad\qquad\qquad$ (kq × nq) $\qquad\qquad\qquad\qquad$ (nq × q)

Let the kq x nq matrix on the right of (7) be R. The matrix R can in turn be written

$$R = \begin{bmatrix} N_1 & N_2 & \cdots & N_k \\ N_1 & N_2 & \cdots & N_k \\ \cdot & \cdot & & \cdot \\ \cdot & \cdot & & \cdot \\ \cdot & \cdot & & \cdot \\ N_1 & N_2 & & N_k \end{bmatrix} \begin{bmatrix} z_1^n \sum_{i=1}^{n} F_i z_1^{-1} & z_1^n \sum_{i=2}^{n} F_i z_1^{-2} & \cdots & F_n z_1 \\ \cdot & & & \\ \cdot & & & \\ \cdot & & & \\ z_k^n \sum_{i=1}^{n} F_i z_k^{-1} & z_k^n \sum_{i=2}^{n} F_i z_k^{-2} & \cdots & F_n z_k \end{bmatrix}$$

or R = NF, under an obvious notation. Suppose $k = rq \geq nq$. Then F will have rank nq. To see this, notice that since F_n is nonsingular, the ultimate set of q columns has rank q. The penultimate set is obtained from the ultimate set by multiplying each set of q rows by z_j^2 and adding to the result $F_{n-1} z_j$. Thus the penultimate set of q columns is linearly independent of the ultimate set. Since the z_j are distinct, such a relationship holds for any two sets of adjacent columns. Thus since the ith set of q columns is not expressible as a linear combination of the sets to its right, and because there are n such sets, the rank of F is nq. The same argument indicates that the k sets of q columns of N are independent, though each has rank 1. Thus rank (N) = k and hence rank (R) $\leq$ min (rank (N), rank (F)) = min (k,nq). The inequality is not strict in general because the k sets of rows of N are identical. Therefore, if k < nq (r<n), there are many solutions to (7); if k = nq (r=n), there is precisely one solution; and if k > nq (r>n), (7) possesses no solution. This completes the proof.

There are several points to be made about this theorem. First, under the Hansen-Sargent configuration, n = m and $F_j = \beta^j G_j$; thus r = n and their solution is unique. Second, the theorem matches that of Blanchard and Kahn (1980), though they use a different method of proof. Appropriately translated, they also make the assumption that det $F_n \neq 0$, remarking that the extension to the alternative case is "straightforward and tedious." A similar claim is appropriate in the present context; one would proceed by partitioning (1) conformably with the nonsingular portion of F_n. Such a procedure has been used by Salemi (1981) for the case $m = n = 1$. Finally, the solutions of Blanchard-Kahn and Salemi involve infinite order matrix convolutions. In addition, in the Blanchard-Kahn solution, these convolutions involve $E_t x_{t+j}$ for $j = 0, 1, \ldots$. Evidently, these convolutions must be simplified in order for the solution to be useful. However, the method of proof used above gives a hint about a representation of the solution which is close enough to a closed form to be useful for estimation in the frequency domain. The nature of this solution is described in the following corollary.

Corollary. Under the conditions of the theorem, the unique solution (3) to (1), when such a solution exists, is given by

$$(8) \qquad y_t = T(L)^{-1} \sum_{j=1}^{nq} \frac{N_j}{L-z_j} \{L^n A(L) - z_j^n A(z_j)\} \varepsilon_t$$

or, if $\{x_t\}$ has an autoregressive representation,

$$(9) \qquad y_t = T(L)^{-1} \sum_{j=1}^{nq} \frac{N_j}{L-z_j} \{L^n I - z_j^n A(z_j) A(L)^{-1}\} x_t.$$

Proof: The polynomial matrix $T(z)^{-1}$ exists everywhere on the open unit disk by corollary 2 to Theorem 1 of the Appendix. Thus, write (5) as

$$(10) \qquad C(z) = T(z)^{-1} \sum_{j=1}^{nq} \frac{N_j}{z-z_j} \{z^n A(z) + z^n \sum_{i=1}^{n-1} \sum_{s=i}^{n-1} F_s C_{s-r} z^{-i} + F_n C_0\}.$$

Then conditions (6) require

(11) $\quad N_j\{z_j{}^n A(z_j) + z_j{}^n \sum_{i=1}^{n-1} \sum_{s=i}^{n-1} F_s C_{s-r} z_j^{-i} + F_n C_0\} = 0 \quad j = 1,\ldots,nq.$

or

$$N_j F_n C_0 = - N_j\{z_j{}^n A(z_j) + z_j{}^n \sum_{i=1}^{n-1} \sum_{s=i}^{n-1} F_s C_{s-r} z_j^{-i}\} \quad j = 1,\ldots,nq.$$

Substituting this expression into (10) and rearranging gives

$$C(z) = T(z)^{-1} \sum_{j=1}^{nq} \frac{N_j}{z-z_j} \{z^n A(z) - z_j{}^n A(z_j) + \sum_{i=1}^{n-1} \sum_{s=i}^{n-1} F_s C_{s-i}(z^{n-i}-z_j^{n-i})\}.$$

The double sum in this expression is of the form

$$T(z)^{-1} \sum_{j=1}^{nq} \frac{N_j}{z-z_j} \{H_1(z^{n-1}-z_j^{n-1}) + H_2(z^{n-2}-z_j^{n-2}) +\ldots+ H_{n-1}(z-z_j)\}$$

where $H_u = \sum_{s=u}^{n-1} F_s C_{s-r}$. But $(z-z_j)$ factors out of each term, leaving

$$T(z)^{-1} \sum_{j=1}^{nq} N_j [H_1(z^{n-2} + z^{n-3} z_j + \ldots + z_j^{n-2}) + \ldots + H_{n-1}]$$

$$= T(z)^{-1}\{[\sum_{j=1}^{nq} N_j][\sum_{v=1}^{n-1} H_v z^{n-v-1}] + [\sum_{j=1}^{nq} z_j N_j][\sum_{v=2}^{n-1} H_v z^{n-v-1}]\}$$

$$+ \ldots + [\sum_{j=1}^{nq} z_j^{n-2} N_j] H_{n-1}$$

$$= 0,$$

since, by Theorem 3 of the Appendix, $\sum_{j=1}^{nq} z_j^\alpha N_j = 0$ for $\alpha = 0,1,\ldots,n-2$.

Thus $C(z)$ is given by

$$C(z) = T(z)^{-1} \sum_{j=1}^{nq} \frac{N_j}{z-z_j} \{z^n A(z) - z_j^n A(z_j)\}$$

which, in operator form, becomes (8), or (9), provided $A(z)^{-1}$ exists on the open unit disk.

To specialize matters somewhat, suppose x_t is characterized by a w-order moving average, so that $A(L) = \sum_{j=0}^{w} A_j L^j$. Then expression (8) becomes

$$y_t = T(L)^{-1} \sum_{j=1}^{nq} N_j \{A(z_j) \sum_{s=0}^{n-1} z_j^{n-1-s} L^s + \sum_{s=0}^{r-1} \sum_{i=s+1}^{r} A_i z_j^{i-s} L^{n+s}\} \varepsilon_t.$$

Alternatively, suppose x_t is characterized by a w-order autoregression, and let $A(L)^{-1} = \sum_{j=0}^{w} B_j L^j$. Then (9) can be written

$$y_t = T(L)^{-1} \sum_{j=1}^{nq} z_j^{n-1} N_j \{B_0 - \sum_{s=1}^{r-1} \sum_{i=s+1}^{r} B_i z_j^{i-s} L^s\} x_t.$$

This is the generalization of Hansen and Sargent's (1981a) formula (51) to <u>ad hoc</u> linear rational expectations models.[1]

The corollary is suggestive in two ways. First, using the bivariate representation theorem of Chapter I, expressions (8) and (3) can be used to form the moving average representation for (x_t, y_t). Estimates of the parameters $F_0, \ldots, F_n$, $G_1, \ldots, G_m$ are obtained by using this moving average representation to form a theoretical spectral density matrix, which is then

---

[1] The two sets of notation do not match exactly. In particular, their $M(z^{-1})$ is $z^{-n} S(z)$. Thus their λ_j corresponds to my z_j, their N_j to my $z_j^{n-2} N_j$, and their $\lambda_j N_j$ to my $z_j^{n-1} N_j$.

used in the frequency-domain maximum likelihood estimation procedure discussed by Hansen and Sargent (1980a). Apparently, the difficult (time consuming) part of such a procedure is the factorization of $z^n(F(z^{-1}) + G(z))$. But the factorization can be done rapidly and accurately using the Smith normal form algorithm described by Pace and Barnett (1974).

Second, though the solution (8) requires $r = n$, a solution resembling (8) can be obtained when $r < n$. Though this solution is only one of a possible infinity of solutions, it illustrates that the algorithm used in the proof of the theorem above is quite general.

To find the solution, factor $z^n(F(z^{-1}) + G(z))$ as $\tilde{S}(z)\tilde{T}(z)$ where det $\tilde{S}(z)$ has exactly n roots, $r < n$ of which lie inside the unit circle. Thus let $z_1, \ldots, z_r$ lie inside the unit circle and let $z_{r+1}, \ldots, z_{nq}$ be the inverses of the roots which lie outside the unit circle. By appropriately multiplying by $\prod_{j=r+1}^{nq} z_j^{-1}$, the partial fractions expansion of $\tilde{S}(z)^{-1}$ can be written

$$\tilde{S}(z)^{-1} = \sum_{j=1}^{nq} \frac{\tilde{N}_j}{z-z_j} .$$

By requiring that (6) hold for $j = 1, \ldots, nq$, one has simply added $nq - rq$ (harmless) restrictions to (7). The arguments in the proof of the corollary go through, leaving

$$y_t = \tilde{T}(L)^{-1} \sum_{j=1}^{nq} \frac{\tilde{N}_j}{L-z_j} \{L^n A(L) - z_j^n A(z_j)\}\varepsilon_t .$$

Beyond providing a simple expression for the solution when one exists, the methods used here are suggestive in still another way - they indicate how to find solutions in models for which neither the theorem above nor the Blanchard-Kahn Theorem seem to apply.

2. An Extension

Though the two theorems are equivalent, the theorem of the previous section is stated quite differently from that of Blanchard-Kahn. By using the law of iterated expectations and appropriately relabeling variables, they would write (1) in the form

$$(12) \quad \begin{bmatrix} Q_{t+1} \\ E_t P_{t+1} \end{bmatrix} = A \begin{bmatrix} Q_t \\ P_t \end{bmatrix} + B X_t.$$

For instance, Q_{t+1} might be $(y_t, y_{t-1}, \ldots, y_{t-m})'$, P_{t+1} might be $(E_t y_{t+1}, E_t y_{t+2}, \ldots, E_t y_{t+n})'$, while A would contain $F_0, F_1, \ldots, F_n, G_1, \ldots, G_m$. They then study the eigenvalues of the matrix A. In other words, they write the polynomial matrix $z^n(F(z^{-1}) + G(z))$ in the form of a linear matrix pencil $zA - I$. Any polynomial matrix can be written in this form (Kailath 1980, Lemma 6.3-20).

As Blanchard and Kahn note, their method sometimes breaks down. For instance, neither it nor the theorem above applies to the model

$$(13) \quad E_{t-1} y_t - P y_t = x_t.$$

Models of this type, termed "withholding equations" in Chapter I, have been studied by Chow (1981). In this case, the matrix polynomial is degenerate: $E_{t-1} y_t = L(L^{-1}(C(L) - C_0))\varepsilon_t$, so that the equation becomes

$$(14) \quad (I-P)C(L)\varepsilon_t = (A(L) + C_0)\varepsilon_t.$$

Thus the appropriate matrix pencil does not involve z, the model cannot be written in the form (12), and the nature of the solution is independent of the eigenvalues of any matrix. But the methods of the previous section are illustrative: the z-transform identity appropriate for (14) is

$$(I-P)C(z) = A(z) + C_0.$$

Then, assuming det $(I-P) \neq 0$,

$$C(z) = (I-P)^{-1}(A(z) + C_0).$$

But unlike the examples of the previous section, this expression is not necessarily consistent. Indeed,

$$C(0) = C_0 = (I-P)^{-1}(A(0) + C_0)$$

which is an identity only if $-PC_0 = A_0$. Then, if P^{-1} exists, the unique solution to (13) is

$$y_t = (I-P)^{-1}(A(L) - P^{-1}A(0))\varepsilon_t.$$

Evidently, the theorem above can be used in conjunction with methods for dealing with withholding equations to find solutions in models of the form

$$(15) \qquad E_{t-w}(\sum_{j=0}^{n} F_j L^{-j} + \sum_{j=1}^{m} G_j L^j) y_t = x_t,$$

where $w > 0$. The class of models (15) and the suggestion given above for dealing with them constitute an answer to the two open questions referred to by Blanchard and Kahn (1980, p. 1309): "The characterization of the class of models not reducible to [12] and the extension of this method to cover models in that class."

Conclusion

A proof of the conditions for existence and uniqueness of solutions to a broad class of vector, linear rational expectations models has been provided.

The solution itself has been derived in the cases for which it exists. In addition, certain models not covered by the theorem have been described, and methods for dealing with such models have been suggested.

Appendix to Chapter IV

In the text, the polynomial matrix $P(z) = z^n(F(z^{-1}) + G(z))$
$= F_n + F_{n-1}z + \ldots + F_0 z^n + G_1 z^{n+1} + \ldots + G_m z^{n+m}$ appears. The following theorem guarantees that any such matrix can be diagonalized.

Theorem 1. Smith Form (Kailath 1980, p. 390).

"For any p by m polynomial matrix $P(z)$ we can find elementary row and column operations, or corresponding unimodular matrices $\{U(z), V(z)\}$, such that

$$U(z) P(z) V(z) = \Lambda(z)$$

where

$$\Lambda(z) = \begin{bmatrix} \lambda_1(z) & & & & \\ & \ddots & & 0 & \\ & & \lambda_r(z) & & \\ \hline & 0 & & 0 \end{bmatrix} \begin{matrix} r \\ \\ p-r \end{matrix}$$
$$\quad\;\; r \qquad\qquad m-r$$

r = the (normal) rank of $P(z)$

and the $\lambda_i(z)$ are unique monic polynomials obeying a <u>division property</u>

$$\lambda_i(z) \mid \lambda_{i+1}(z) \qquad i = 1, \ldots, r-1.$$

Moreover, if we define

$\Delta_i(z)$ = the gcd [greatest common divisor] of all i x i minors of $P(z)$

then we can identify

$$\lambda_i(z) = \frac{\Delta_i(z)}{\Delta_{i-1}(z)}, \qquad \Delta_0(z) = 1.$$

The matrix $\Lambda(z)$ is called the <u>Smith form</u> of $P(z)$."

<u>Proof</u>: Kailath (1980, p. 391).

Remark: A unimodular matrix $U(z)$ is a square matrix whose determinant is a nonzero constant. The normal rank of a matrix $P(z)$ is its rank for almost all z. Thus if $P(z)$ is square and $\det P(z) = c \prod_{j=1}^{k} (z-z_j)$, the normal rank of $P(z)$ is the rank of $P(z)$ for all $z \neq z_j$, $j = 1,\ldots,k$. A monic polynomial $\lambda(z)$ is one with a unit coefficient on the highest power of z. Finally, the division property means that

$$\lambda_{i+1}(z) = \lambda_i(z)\gamma_i(z),$$

i.e., that $\lambda_i(z)$ divides $\lambda_{i+1}(z)$ without remainder.

<u>Corollary 1</u>. If $P(z)$ is a $(q \times q)$ matrix whose determinant is nonzero on the unit circle and $P(0)$ is nonsingular, then $P(z)$ can be written as

$$P(z) = \tilde{U}(z) \, \tilde{V}(z)$$

where the roots of $\det \tilde{U}(z)$ are inside the unit circle while those of $\tilde{V}(z)$ are outside the unit circle.

<u>Proof</u>: Since $P(0)$ is nonsingular, the normal rank of $P(z)$ is q. Then the Smith form of $P(z)$ is

$$\Lambda(z) = \text{diag } (\lambda_1(z),\ldots,\lambda_q(z)).$$

Now factor each of the polynomials $\lambda_i(z)$ and write

$$\Lambda(z) = \text{diag } (\lambda_1^1(z),\ldots,\lambda_q^1(z)) \text{ diag } (\lambda_1^2(z),\ldots,\lambda_q^2(z))$$

where, for each j, the roots of $\lambda_j^1(z)$ are inside, and those of $\lambda_j^2(z)$ are outside, the unit circle. Then write $P(z) = U(z)^{-1} \Lambda(z) V(z)^{-1}$ and let

$$\tilde{U}(z) = U(z)^{-1} \text{ diag } (\lambda_1^1(z),\ldots,\lambda_q^1(z))$$

$$\tilde{V}(z) = \text{diag } (\lambda_1^2(z),\ldots,\lambda_q^2(z)) \, V(z)^{-1}.$$

Corollary 2. If the roots of det $T(z)$ are all outside the unit circle, $T(z)^{-1}$ exists on the open unit disk.

Proof: Let the Smith form of $T(z)$ be $\Lambda(z)$. Then det $T(z)$ is proportional to det $\Lambda(z)$. The roots of $\Lambda(z)$ are all outside the unit circle. Thus

$$\Lambda(z)^{-1} = \text{diag } (\lambda_1^{-1}(z),\ldots,\lambda_q^{-1}(z))$$

exists on the unit disk, as does

$$T(z)^{-1} = V(z) \, \Lambda(z)^{-1} \, U(z)$$

where $U(z)$ and $V(z)$ are unimodular.

The following theorem and corollary establish that the rank of each of the matrices N_j in (5) is 1.

Theorem 2. Let the (qxq) matrix $H(z) = \sum_{j=0}^{n} H_j z^j$ and let the nq roots of det $H(z) = h \sum_{k=1}^{nq} (z-z_k)$ be distinct. Let the normal rank of $H(z)$ be q. Then rank $(H(z_k)) = q-1$, $k=1,\ldots,nq$.

Proof: Reduce $H(z)$ to Smith form:

$$U(z) \, H(z) \, V(z) = \text{diag } (\lambda_1(z),\ldots,\lambda_q(z)).$$

Now

$$\det (U(z) \, H(z) \, V(z)) = h' \prod_{k=1}^{nq} (z-z_k)$$

since det $U(z)$ and det $V(z)$ are constants. Also,

$$\det(\operatorname{diag}(\lambda_1(z),\ldots,\lambda_q(z))) = \prod_{r=1}^{q} \lambda_r(z).$$

Thus the functions $h' \prod(z-z_k)$ and $\prod \lambda_r(z)$ have the same nq unique zeros. Without loss of generality, assume z_k is a zero of $\lambda_q(z)$ only. Then

$$U(z_k) H(z_k) V(z_k) = \operatorname{diag}(\lambda_1(z_k),\ldots,\lambda_{q-1}(z_k),0).$$

Thus $\operatorname{rank}(H(z_k)) = \operatorname{rank}(V(z_k) H(z_k) V(z_k)) = q-1$.

Corollary. $\operatorname{adj} H(z_k)$ has unit rank.

Proof: Without loss of generality, assume that the first $q-1$ columns of $H(z_k)$ are linearly independent. Then the qth column, $H^q(z_k)$, is uniquely expressed as a linear combination of the first $q-1$:

$$H^q(z_k) = \sum_{j=1}^{q-1} a_j H^j(z_k).$$

Thus the set of $(q \times 1)$ vectors y such that $H(z_k)y = 0$ are of the form $a_0(a_1, a_2, \ldots, a_{q-1}, -1)$. Each column of $\operatorname{adj} H(z_k)$ is such a vector, since $H(z_k) \operatorname{adj} H(z_k) = 0_{qq}$. Thus the columns of $\operatorname{adj} H(z_k)$ are proportional. Thus $\operatorname{rank}(\operatorname{adj} H(z_k)) \leq 1$. But $\operatorname{adj} H(z_k)$ is not the zero matrix. Thus $\operatorname{rank}(\operatorname{adj} H(z_k)) = 1$.

Theorem 3 establishes the conditions on N_j which were used to obtain the solution (8).

Theorem 3. Let $S(z)$ be a $(q \times q)$ polynomial matrix of order $r > 0$:

$$S(z) = \sum_{j=0}^{r} S_j z^j$$

with $\det S(z) = \sum_{j=0}^{rq} d_j z^j = d \prod_{j=1}^{rq}(z-z_j)$. Suppose the roots are distinct. Write the partial fractions expansion of $S(z)^{-1}$, which exists almost everywhere in the complex plane, as

$$S(z)^{-1} = \sum_{j=1}^{rq} \frac{N_j}{z-z_j}.$$

Then $\sum_{j=1}^{rq} z_j^{\alpha} N_j = 0(q \times q)$ for $\alpha = 0,1,\ldots,r-2$.

Proof: First, write $S(z)^{-1}$ as

$$S(z)^{-1} = \frac{\text{Adj } S(z)}{\det S(z)}.$$

Then $(\det S(z)) I = (\text{Adj } S(z)) S(z)$ is a polynomial matrix of order rq. Thus, since $S(z)$ is of order r, Adj $S(z)$ must be of order $rq - r$:

$$\text{Adj } S(z) = \sum_{j=1}^{rq-r} S_j^* z^j.$$

Next, note that

$$\text{Adj } S(z) = (\det S(z)) S(z)^{-1}$$

or

$$\sum_{j=1}^{rq-r} S_j^* z^j = (\det S(z)) \sum_{j=1}^{rq} \frac{N_j}{z-z_j}$$

(A1) $$\sum_{j=1}^{rq-r} S_j^* z^j = d \sum_{j=1}^{rq} \frac{\prod_{i=1}^{rq}(z-z_i)}{z-z_j} N_j.$$

Since $\prod(z-z_j)$ is of order rq, the polynomial matrix on the right-hand side of (A1) is of order $rq-1$, while the order of the matrix on the left is $rq-r$. If

(A1) is to hold, the coefficients on z^{rq-1}, z^{rq-2},...,z^{rq-r+1} on the right-hand side of (A1) must equal zero. Write

$$\prod_{i=1}^{rq}(z-z_j) = z^{rq} + P_1 z^{rq-1} + \ldots + P_{rq}.$$

From Gantmacher (1959, p. 87) or Chrystal (1893, p. 437), P_u is given by Newton's formulas

$$P_u = -\frac{1}{u}(s_u + P_1 s_{u-1} + \ldots + P_{u-1} s_1) \qquad u = 1,\ldots,rq$$

where $s_u = \sum_{j=1}^{rq} z_j^u$. Following Chrystal, write

(A2) $$\frac{\prod_{i=1}^{rq}(z-z_j)}{z-z_j} = z^{rq-1} + {}_j P_1 z^{rq-2} + \ldots + {}_j P_{rq-1}.$$

The formulas for ${}_j P_u$ are given by Chrystal as

(A3) $$\quad {}_j P_u = z_j^u + P_1 z_j^{u-1} + \ldots + P_u.$$

Thus the coefficient on z^{rq-1} on the right-hand side of (A1) is $d \sum_{j=1}^{rq} N_j$.

Since $d \neq 0$ and $Adj(S(z))$ is only of order $rq-r$, $\sum_{j=1}^{rq} N_j = 0$. Now the coefficient on z^{rq-2}, which must equal zero, is

$$d \sum_{j=1}^{rq} {}_j P_1 N_j = d \sum_{j=1}^{rq}(z_j - \sum_{j=1}^{rq} z_j) N_j$$

$$= d \sum_{j=1}^{rq} z_j N_j - d \sum_{j=1}^{rq}(\sum_{j=1}^{rq} z_j) N_j$$

$$= d \sum_{j=1}^{rq} z_j N_j$$

since $\sum_{j=1}^{rq} N_j = 0$ from above. Thus $\sum_{j=1}^{rq} z_j N_j = 0$. It is clear from (A3) that such a relationship holds for the coefficients on $z^{rq-3},\ldots,z^{rq-r+1}$; i.e.,

$$\sum_{j=1}^{rq} z_j^{\alpha} N_j = 0 \text{ for } \alpha = 0,1,\ldots,r-2.$$

V

A Set of Optimization Examples

One of the advantages of the technique studied in previous chapters is that it can be used to solve a wide class of rational expectations models. The technique is useful whether the expectational difference equation is viewed as the primitive objective of the model or as the first-order condition of some fully specified optimization problem. The purpose of this chapter is to investigate this latter claim; to show how expectational difference equations arise as necessary conditions for equilibria in a class of dynamic equilibrium models, and to show how the equilibria can be calculated using the techniques decribed in Chapters I-IV.

The first-order conditions for a quadratic, stochastic dynamic optimization problem are derived in section 1 below. The resulting expectational difference equation is solved under three sets of assumptions concerning the environment faced by the economic agents in sections 2-4. Section 2 studies the simplest rational expectations equilibrium. Section 3 treats a problem in which the self-fulfilling nature of rational expectations is important. In the model studied in section 4, the competitive equilibrium is not efficient. The calculations there illustrate the advantage the expectational difference equation techniques can have over conventional, iterative methods.

1. The Model

Each of a large, fixed number n of identical firms which behave competitively produces output y_t using capital k_t according to the linear production technology $y_t = f_0 k_t$. Firms rent capital at the rate w_t and, at date t, each bears real, internal costs of $(d/2)(k_t - k_{t-1})^2$ of having adjusted its capital stock. The firms seek to maximize discounted net cash flow; each firm, by choice of $\{k_t\}$ seeks to maximize

(1) $$E_0 \sum_{t=0}^{\infty} \beta^t \{p_t(f_0 k_t) - w_t k_t - (d/2)(k_t - k_{t-1})^2\}$$

where β is the discount rate, p_t is the price of output, and $E_0(x)$ denotes the linear least squares forecast of x based upon information available at time 0. The industry faces a demand curve for its output given by

(2) $$p_t = D_0 - D_1 Y_t + u_t,$$

where $Y_t = n y_t$ is industry output, D_0 and D_1 are nonnegative parameters, and $\{u_t\}$ can be taken to be a linearly regular covariance stationary, zero-mean stochastic process (LRCSSP^0). In addition, the industry faces a factor supply curve of the form

(3) $$w_t = S_0 + S_1 K_t + S_2 K_{t-1} + v_t,$$

where $K_t = n k_t$ is the industry-wide capital stock, S_0 and S_1 are nonnegative, and v_t is a LRCSSP^0. The factor supply schedule permits feedback ($S_2 \neq 0$) from the market-wide capital stock to the rental rate.[1] In this case, the aggregate of the firms' decision variables $\{K_t\}$, Granger-causes $\{w_t\}$.[2] Of course, individual firms, behaving competitively, take $\{p_t\}$ and $\{w_t\}$ parametrically.

To obtain the representative firm's first-order condition as an expectational difference equation, it is convenient to define the net cash flow at time t as

(4) $\qquad R(k_t, k_{t-1}, p_t, w_t) = p_t f_0 k_t - w_t k_t - \frac{d}{2}(k_t - k_{t-1})^2.$

Thus the problem (1) becomes: maximize

(5) $\qquad E_0 \sum_{t=0}^{\infty} \beta^t R(k_t, k_{t-1}, p_t, w_t)$

by choice of $\{k_t\}$, where the initial capital stock, k_{-1}, is taken as given. Using Leibniz's rule to differentiate (5) with respect to $k_0, k_1, k_2, \ldots$, one obtains

$$\beta^t E_0 \{R_1(k_t, k_{t-1}, p_t, w_t) + \beta R_2(k_{t+1}, k_t, p_{t+1}, w_{t+1})\} = 0 \qquad t=0,1,\ldots,$$

where $R_i(\ )$ is the partial derivative of R with respect to its ith argument. Using the law of iterated expectations, the condition becomes

$$\beta^t E_0 \{R_1(k_t, k_{t-1}, p_t, w_t) + \beta E_t R_2(k_{t+1}, k_t, p_{t+1}, w_{t+1})\} = 0 \qquad t=0,1,\ldots.$$

If this equation is to hold for all realizations of (p_t, w_t), the terms in braces must equal zero:

---

[1] In Sargent's (1980) analysis of this model, the once-lagged wage, w_{t-1}, was allowed to enter (3) directly. The dependence could be maintained here, but it would not add much, and it would make the important results to be derived below somewhat less transparent.

[2] This case is apparently ruled out in Hansen and Sargent (1980a).

(6) $\quad E_t R_2(k_{t+1}, k_t, p_{t+1}, w_{t+1}) + \frac{1}{\beta} R_1(k_t, k_{t-1}, p_t, w_t) = 0 \quad t=0,1,\ldots$

The advantage of a quadratic return function R is readily apparent from (6): the first-order conditions are then linear expectational difference equations.[3] Calculating $R_1(\ )$ and $R_2(\ )$ from (4), expression (6) becomes

$$E_t d(k_{t+1}-k_t) + \frac{1}{\beta}(p_t f_0 - w_t - d(k_t - d_{t-1})) = 0 \quad t=0,1,\ldots,$$

or

(7) $\quad E_t k_{t+1} - (1+\beta^{-1})k_t + \beta^{-1} k_{t-1} = (d\beta)^{-1}(w_t - f_0 p_t) \quad t=0,1,\ldots,$

where it has been assumed that firms know k_t at time t.

In the next several sections, equation (7) is solved using the methods of Chapters I-IV given various conditions on $\{w_t\}$ and $\{p_t\}$, and given one crucial proviso: to ensure that the infinite-horizon present value calculation is meaningful, the condition

(8) $\quad \lim_{t\to\infty} \beta^t E_0 k_t^2 = 0$

will be imposed.[4] This condition means that capital may grow, but not "too rapidly." More formally, rather than requiring that the z-transform of the moving average coefficients of $\{k_t\}$ be analytic on the open unit disk, (8) requires only that the function be analytic on the disk with radius $\sqrt{\beta}$. Of course, if $\beta = 1$, the analyticity requirement for this problem exactly matches that of Chapters I-IV.

---

[3]This point is made in some detail by Lucas and Sargent (1981).

[4]See Hansen and Sargent (1981a).

2. Perfectly Elastic Supply and Demand

Assume $D_1 = S_1 = S_2 = 0$, so that firms face perfectly elastic industry-wide supply and demand curves. This case, with the additional convenience assumption $D_0 = S_0 = 0$, is the one studied by Hansen and Sargent (1980a).

To find the decision rules firms will use, it is necessary to specify how the prices p_t and w_t evolve, as well as the information the firms have about these processes. For reasons which will become clear in the next section, it is convenient to specify this information in terms of the demand and supply "shift" variables u_t and v_t. Of course, with $D_0 = D_1 = S_0 = S_1 = S_2$, $\{p_t\} = \{u_t\}$ and $\{w_t\} = \{v_t\}$. Thus suppose that the jointly stationary processes $\{u_t\}$ and $\{v_t\}$ have the Wold moving average representation

$$(9) \quad \begin{bmatrix} u_t \\ v_t \end{bmatrix} = \begin{bmatrix} A_1(L) & B_1(L) \\ A_2(L) & B_2(L) \end{bmatrix} \begin{bmatrix} \varepsilon_{1t} \\ \varepsilon_{2t} \end{bmatrix} \equiv R(L)\varepsilon_t,$$

that firms know these laws of motion, and that, at date t, firms' information sets include $(u_t, u_{t-1}, \ldots, v_t, v_{t-1}, \ldots)$. As in Chapter I, it will be assumed that $(u_t \ v_t)'$ has a vector autoregressive representation, so that knowledge of current and past prices is equivalent to knowledge of current and past innovations $(\varepsilon_{1t} \ \varepsilon_{2t})'$.

A solution $\{k_t\}$ to (7) which satisfies (8) will be sought in the space spanned by current and past values of the (vector) driving process $(u_t \ v_t)'$.[5] Thus assume that k_t can be written

$$(10) \quad k_t = C(L)\varepsilon_{1t} + D(L)\varepsilon_{2t}.$$

---

[5] As in Chapter I, the trivariate system $(k_t \ u_t \ v_t)'$ could be assumed to be nonsingular.

Expressions (9) and (10) can be used in (7) to obtain the z-transform identities

$$C(z) = (d\beta)^{-1}\{(1-z)(1-\beta^{-1}z)\}^{-1} \{z(A_2(z)-f_0A_1(z)) + d\beta C_0\}$$
$$D(z) = (d\beta)^{-1}\{(1-z)(1-\beta^{-1}z)\}^{-1} \{z(B_2(z)-f_0B_1(z)) + d\beta D_0\}.$$

For $C(z)$ and $D(z)$ to be analytic for $|z| < \sqrt{\beta}$, C_0 and D_0 must be set so that the residues at $z = \beta$ are zero:

$$\beta(A_2(\beta) - f_0A_1(\beta)) + d\beta C_0 = 0$$
$$\beta(B_2(\beta) - f_0B_1(\beta)) + d\beta D_0 = 0.$$

With C_0 and D_0 set in this way, the unique solution is given by

(11) $\quad k_t = (d\beta)^{-1}\{(1-L)(1-\beta^{-1}L)\}^{-1}$

$$\times \{[LA_2(L) - \beta A_2(\beta) - f_0(LA_1(L) - \beta A_1(\beta))]\varepsilon_{1t}$$
$$+ [LB_2(L) - \beta B_2(\beta) - f_0(LB_1(L) - \beta B_1(\beta))]\varepsilon_{2t}\}.$$

Define the operator

$$K(A(L),\rho^{-1}) = LA(L) - \rho^{-1}A(\rho^{-1}).$$

Then (11) can be written (recall $p_t = u_t$ and $w_t = v_t$)

(12) $\quad k_t = k_{t-1} + (d\beta)^{-1}\{1-\beta^{-1}L\}^{-1}$

$$\times \{K(A_2(L),\beta)-f_0K(A_1(L),\beta) \quad K(B_2(L),\beta)-f_0K(B_1(L),\beta)\}R(L)^{-1}(p_t \ w_t)'.$$

This expression gives the capital accumulation plan of an individual firm as a function of the currently prevailing output and factor prices as well as (possibly) the entire histories of these prices. Furthermore, the coefficients on current and past prices in this investment function are not independent of the way--R(L)--the prices evolve: (12) clearly exhibits the cross-equation restrictions of rational expectations.

An interesting special case occurs when R(L) is diagonal--$A_2(L) = B_1(L) = 0$. In this situation neither variable helps predict (Granger-causes) the other. If, in addition, u_t and v_t can be taken to be first order Markov processes ($A_1(L) = (1-aL)^{-1}$, $B_2(L) = (1-bL)^{-1}$), expression (12) simplifies substantially:

(13) $\quad k_t = k_{t-1} + d^{-1}\{[f_0/(1-a/\beta)]p_t - [1/(1-b/\beta)]w_t\}$.

There are, of course, other methods of obtaining (12) or (13). One, discussed in Hansen and Sargent (1980a), exploits the quadratic nature of (1) and invokes the certainty-equivalence principle. In this procedure, $\{p_t\}$ and $\{w_t\}$ are treated as known until after the deterministic version of (7) is solved to give k_t in terms of future wages and prices; the Wiener-Kolmogorov formula is then used to compute the conditional expectations of these future values. Another method, pursued in Hansen and Sargent (1981a), involves recasting the problem of maximizing (1) subject to (2) and (3) in terms of the "optimal linear regulator problem." This method allows the decision rule to be computed by an iterative technique, long known to engineers, involving the matrix Riccati difference equation.

3. Inelastic Demand; Perfectly Elastic Supply

Suppose $D_1 > 0$, so that the industry faces a downward sloping demand curve. The convenience assumption of no intercepts ($D_0=S_0=0$) and the assumption that the factor is supplied elastically ($S_1=S_2=0$) will be maintained. The downward sloping demand curve makes the nature of the equilibrium somewhat more interesting than that of the previous section; in that model, there was an "exogenous truth" about $\{p_t\}$ which firms discovered, but with $D_1 > 0$ there is an "endogenous truth" being produced in equilibrium.

To see this, notice that firms decide on k_t based on forecasts of prices. But those prices are affected via (2) by the aggregate of firms' decisions, $K_t = nk_t$. In equilibrium, the forecasts must on average be correct. The analysis of previous chapters (especially Chapter II, section 6) makes it clear that this equilibrium is a fixed point (in the space of analytic functions) which corresponds to the moving average representation of a stochastic process.

Following Sargent (1979b), the equilibrium can be defined as follows. A <u>rational expectations equilibrium</u> is a pair of stochastic processes $\{p_t\}_{t=0}^{\infty}$ and $\{k_t\}_{t=0}^{\infty}$ which satisfies two conditions:

(i) Given firms' decision rules for $\{k_t\}_{t=0}^{\infty}$, the stochastic process for prices, $\{p_t\}_{t=0}^{\infty}$ clears the output market; i.e., (2) holds.

(ii) When firms take the stochastic process for $\{p_t\}_{t=0}^{\infty}$ as given, the process $\{k_t\}_{t=0}^{\infty}$ maximizes discounted net cash flow (1).

To compute the equilibrium, substitute the supply equation (2) into the first-order condition (7) to obtain

$$E_t k_{t+1} - (1+\beta^{-1})k_t + \beta^{-1}k_{t-1} = (d\beta)^{-1}(w_t + f_0(D_1 y_t - u_t))$$
$$= (d\beta)^{-1}(w_t + f_0(D_1 n f_0 k_t - u_t))$$

or

(14) $E_t k_{t+1} - (1+\beta^{-1} + (d\beta)^{-1} n f_0^2 D_1) k_t + \beta^{-1} k_{t-1} = (d\beta)^{-1}(w_t - f_0 u_t).$

Equation (14) is a version of (7) in which the demand shift variable u_t stands in for p_t (here, $u_t \neq p_t$) and in which the coefficient on k_t has changed. Writing

$$\rho_1 + \rho_2 = 1 + \beta^{-1} + (d\beta)^{-1} n f_0^2 D_1$$
$$\rho_1 \rho_2 = \beta^{-1},$$

it is clear that the introduction of a downward-sloping demand curve changes the precise position of the roots of the characteristic equation of the system, but not the general location; one root remains inside the "β" circle, the other remains outside. So long as $D_1 \geq 0$, (14) inherits the properties of (7); the solution (equilibrium) exists and is unique. Using the calculations of the previous section, this solution is given by

(15) $\quad k_t = \rho_1 k_{t-1} + (d\beta)^{-1} \{1-\rho_2 L\}^{-1}$

$\qquad \times \{K(A_2(L),\rho_2^{-1}) - f_0 K(A_1(L),\rho_2^{-1}) \quad K(B_2(L)(,\rho_2^{-1}) - f_0 K(B_1(L),\rho_2^{-1})\}$

$\qquad \times R(L)^{-1} (u_t \quad v_t)'.$

Expression (15) describes the per-firm capital stock and thus exhibits properties similar to those of the decision rule (12) of the previous section. In particular, the equilibrium law of motion of capital depends upon the way output and factor prices evolve.

Provided some care is taken in its application, the certainty-equivalence procedure alluded to in section 2 can also be used to calculate (15). In addition, the iterative, matrix-Riccati techniques are available. But the implementation of either of these methods is simplified if the problem of this section can be rewritten in a form which more closely resembles that of section 2. That is, can one formulate a problem in which the equilibrium (15) is the decision rule for some agent? The affirmative answer to this question, proposed in a related context by Lucas and Prescott (1971), stems from the fact that the competitive equilibrium in this model is efficient, and thus maximizes the welfare of a fictitious "social planner." Thus consider the objective of a planner who seeks to maximize discounted consumer surplus less costs of production: the problem is to maximize

(16) $\quad E_0 \sum_{t=0}^{\infty} \beta^t \{ \int_0^{Y_t} (D_0 - D_1 x + u_t) dx - w_t k_t - (d/2)(k_t - k_{t-1})^2 \}$

subject to the law of motion (3) for $\{w_t\}$. Upon substituting $D_0 = 0$ and $Y_t = nf_0 k_t$ into (16), it is easily verified that the first-order condition for this problem is (14). By taking advantage of the efficiency of competitive equilibrium, it is thus possible to rewrite the equilibrium problem to make certainty-equivalent or matrix-Ricatti methods straightforwardly applicable. The next section introduces conditions under which the rewriting procedure is not possible.

4. Feedback from Decisions to States

Suppose now that $D_0 = S_0 = S_1 = 0$ but that $S_2 \neq 0$. In this case, the demand curve facing the industry is not perfectly elastic and the factor supply equation admits dependence of the current factor price on previous decisions of agents. As in the model studied by Kydland and Prescott (1977), "in effect this produces an externality." To see why, notice that if $S_2 > 0$, larger current capital stocks mean higher future rental rates. A social planner would take this into account in a way that the competitive equilibrium does not.[6] Thus the competitive equilibrium, defined above, is not optimal and therefore cannot be calculated by the straightforward application of certainty-equivalent or matrix-Riccati methods to a social planning problem. But there is a procedure available to calculate the equilbrium; Sargent (1980) labels it "the Kydland-Prescott method," as they used it in a slightly different context in their 1977 paper.

---

[6]Notice that the parameter S_2 is being treated as structural. In particular, the optimization problem for which (3) is a decision rule is not specified.

To see how the method works, consider a finite horizon version of (1): firms maximize

$$(17) \quad E_0 \sum_{t=0}^{T} \beta^t \{p_t f_0 k_t - w_t k_t - (d/2)(k_t - k_{t-1})^2\}$$

subject to (2), (3), and k_{-1} given. To do this, the firms must forecast future prices. But because of (3), price forecasts will involve forecasts of how the industry-wide capital stock evolves. Thus, assuming for convenience that u_t and v_t are independent first order Markov processes (as in (13)) and adapting the notation in Sargent (1980), suppose that at time t, firms believe that $\{k_t\}$ evolves according to[7]

$$(18) \quad K_t = h_1^{t,T} K_{t-1} + h_2^{t,T} v_t + h_3^{t,T} u_t,$$

where notice that the coefficients depend not only on the date at which the perception is made, but also the planning horizon. The reason for this latter dependence will become clear presently.

The solution to the problem of maximizing (17) subject to (2), (3), k_{-1}, and the perception (18) is a sequence of contingency plans for setting k_t as a function of k_{t-1}, K_{t-1}, u_t, and v_t:

$$(19) \quad k_t = d_1^{t,T} k_{t-1} + d_2^{t,T} v_t + d_3^{t,T} u_t + d_4^{t,T} K_{t-1} \quad t=0,1,\ldots,T.$$

In equilibrium, firms' perceptions must turn out to be correct. Thus

$$h_1^{t,T} = d_1^{t,T} + n d_4^{t,T}$$
$$h_2^{t,T} = n d_2^{t,T}$$
$$h_3^{t,T} = n d_3^{t,T}$$

---

[7] Only current shocks appear in (18) by virtue of the first-order Markov assumption.

The reason for the dependence of (18) upon the length of the horizon should now be clear: the horizon length influences the nature of the contingency plans (19) which must, in equilibrium, aggregate to produce (18).

The Kydland-Prescott algorithm calculates the competitive, rational expectations equilibrium law of motion for this model as the limit of the "first period" (time 0) functions in (18) and (19) as the horizon becomes infinite:

$$d_i = \lim_{T \to \infty} d_i^{0,T} \qquad i = 1,\ldots,4$$

$$h_i = \lim_{T \to \infty} h_i^{0,T} \qquad i = 1,\ldots,3.$$

The iterations (on a large-dimension difference equation) to do this are described in some detail in Sargent (1980). Under the conditions of this problem, these iterations are known (sections 3 and 4 above) to converge if $S_2 = 0$. But if feedback is allowed, all that is known is that if the process converges, it converges to an equilibrium.

The introduction of feedback from aggregates of decisions to state variables complicates the iterative procedures--it changes the problem from one for which convergence conditions are known to one for which they are not. But $S_2 \neq 0$ does not make the appropriate expectational difference equation more difficult to analyze. To see this, substitute the appropriate versions of (2) and (3) into (7) to obtain

$$E_t k_{t+1} - (1+\beta^{-1})k_t + \beta^{-1}k_{t-1} = (d\beta)^{-1}\{S_2 K_{t-1} + v_t + f_0(D_1 Y_t - u_t)\}$$
$$= (d\beta)^{-1}\{nS_2 k_{t-1} + v_t + nf_0^2 D_1 k_t - f_0 u_t\}$$

or

(20) $\quad E_t k_{t+1} - (1+\beta^{-1}+(d\beta)^{-1}nf_0^2 D_1)k_t + (\beta^{-1}-(d\beta)^{-1}nS_2)k_{t-1} = v_t - f_0 u_t.$

Writing

$$\tilde{\rho}_1 + \tilde{\rho}_2 = 1 + \beta^{-1} + (d\beta)^{-1} n f_0^2 D_1$$
$$\tilde{\rho}_1 \tilde{\rho}_2 = \beta^{-1} - (d\beta)^{-1} n S_2,$$

it is clear that the introduction of feedback does not change the sum of the factors in the characteristic equation ($\tilde{\rho}_1 + \tilde{\rho}_2 = \rho_1 + \rho_2$), but it does change the product ($\tilde{\rho}_1 \tilde{\rho}_2 \neq \rho_1 \rho_2$). Thus unlike inelastic demand, the presence of capital stock to rental rate feedback can affect the existence of equilibrium. The reason for this is that with $S_2 \neq 0$, $1/\sqrt{\beta} < |\tilde{\rho}_1|, |\tilde{\rho}_2|$ is possible. Multiple equilibria, on the other hand, cannot occur: so long as $D_1 \geq 0$, the factors must sum to at least $1 + \beta^{-1}$; both cannot therefore have modulus less than $1/\sqrt{\beta}$. Of course, if the equilibrium exists--if $|\tilde{\rho}_1| < 1/\sqrt{\beta} < |\tilde{\rho}_2|$--it is given <u>in-closed form</u> by (15) with ρ_1 and ρ_2 replaced by $\tilde{\rho}_1$ and $\tilde{\rho}_2$.

A numerical example will help illustrate the advantage the closed form (15) has over the iterative techniques. The parameter values are $d=1$, $\beta=0.9$, $D_1=10^{-3}$, $n=10^3$, $f_0=2$, $A_2(L)=B_1(L)=0$, $A_1(L)=(1-0.7L)^{-1}$, $B_2(L)=1$, and $D_0=S_0=S_1=0$. For these values, $\tilde{\rho}_1$ and $\tilde{\rho}_2$ are given by

$$\tilde{\rho}_1, \tilde{\rho}_2 = (59/18) \pm 0.5[(59^2/9^2) - (40/9)(1-1000S_2)]^{\frac{1}{2}}.$$

The region for the existence ordering $|\tilde{\rho}_1| < 1/\sqrt{\beta} < |\tilde{\rho}_2|$ is

$$-0.0042191461 < S_2 < 0.008219146.$$

Thus suppose that $S_2 = 0.008$. In this case, $\tilde{\rho}_1 = -1.03 > -1/\sqrt{0.9} = -1.05$, and the industry experiences discount rate-damped oscillations--cycles. The equilibrium can be calculated (via (15)) quite simply by hand, and it takes much less than one second of computer time on the CYBER 172 at the University of Minnesota. The calculation of equilibrium using the Kydland-Prescott algorithm, on the other hand, took 347 iterations and 10 seconds.[8] At current rates, the calculation costs about $1. Alternatively, suppose $S_2 = 0.0085$. In this case, $\tilde{\rho}_1 = -1.09 < -1/\sqrt{\beta}$, and the industry starts off on a path of explosive oscillations. Factoring the polynomial in (20), it is readily apparent that an equilibrium does not exist. But using the iterative scheme, nonexistence is evident only when the scheme "explodes;" in this case, calculations were terminated at the (arbitrary) limit of 2000 iterations after 60 seconds and about $6.

The example illustrates the value of (15) even if the equilibrium is to be calculated only one time. But the advantage of a closed form is even more evident in estimation contexts in which each of a possibly large number of evaluations of the likelihood function requires computation of an equilibrium.

It is the effect of S_2 on the product of the factors in (20) that influences the existence of equilibrium. But a similar phenomenon will occur whenever the coefficient on z^2 in the identities associated with (7) differs from β^{-1}. Thus, for example, the presence of Y_{t-1} in the demand function (2) could give rise to nonexistence examples similar to the one studied above.

Conclusion

In addition to being useful for solving rational expectations models postulated at the level of supply and demand equations, the methods of Chapters I-IV can be used to compute equilibria for models in which agents optimize, markets clear at each date, and expectations are fulfilled. The

---

[8] The convergence criterion was 10^{-5}. That is, convergence was assumed if $|d_i^{0,T+1} - d_i^{0,T}| < 10^{-5}$. The calculations were performed using the Kydland-Prescott algorithm described by Sargent (1980). The program was written and run by Ian Bain at the Federal Reserve Bank of Minneapolis.

methods work whether or not the equilibrium is optimal; they can be used to calculate a competitive equilibrium or, if it is different, the "social planning" equilibrium. Furthermore, the methods make possible a potentially very large gain in computational efficiency over iterative techniques.

References

Aoki, Masanao and Mathew Canzoneri. (1979) "Reduced Forms of Rational Expectations Models." *Quarterly Journal of Economics*, 93 (February), 59-71.

Black, Fischer. (1974) "Uniqueness of the Price Level in Monetary Growth Models with Rational Expectations." *Journal of Economic Theory*, 7 (January), 53-65.

Blanchard, Olivier J. (1978) "The Solution of Linear Difference Models Under Rational Expectations: Its Application to the Hahn Problem." Harvard Harvard Institute of Economic Research Discussion Paper No. 628 (June).

_____. (1979) "Backward and Forward Solutions for Economies with Rational Expectations." *American Economic Review*, 69 (May), 114-118.

_____ and C.M. Kahn. (1980) "The Solution of Linear Difference Models Under Rational Expectations." *Econometrica*, 48 (July), 1305-1311.

Brock, William A. (1975) "A Simple Perfect Foresight Monetary Model." *Journal of Monetary Economics*, 1 (April), 133-150.

Burmeister, Edwin. (1980) "On Some Conceptual Issues in Rational Expectations Modeling." *Journal of Money, Credit, and Banking*, 12 part 2 (November), 800-816.

Cagan, Phillip. (1956) "The Monetary Dynamics of Hyperinflation." *Studies in the Quantity Theory of Money*. Edited by M. Friedman. Chicago: University of Chicago Press.

Chow, Gregory C. (1981) "Solution and Estimation of Simultaneous Equations Under Rational Expectations," Princeton Econometric Research Memorandum No. 291 (October).

Christiano, Lawrence J. (1980) "Rational Expectations, Hyperinflation, and the Demand for Money." Federal Reserve Bank of Minneapolis Working Paper No. 163 (November).

Chrystal, G. (1893) <u>Algebra</u> Part I, third edition. London: Adam and Charles Black.

Churchill, Ruel V.; Brown, James W.; and Roger F. Verhey. (1974) <u>Complex Variables and Applications</u>. New York: McGraw-Hill.

de Branges, Louis and James Rovnyak. (1966) <u>Square Summable Power Series</u>. New York: Holt, Rinehart, and Winston.

Dornbusch, Rudiger. (1976) "Expectations and Exchange Rate Dynamics." <u>Journal of Political Economy</u>, 84 (December), 1161-1176.

Eckstein, Zvi. (1980) "Formulating and Estimating Dynamic Models of Agricultural Production and Land Allocation: The Egyptian Case." Unpublished Ph.D. dissertation, the University of Minnesota.

Eichenbaum, Martin S. (1981) "A Rational Expectations Equilibrium Model of the Cyclical Behavior of Employment and Inventories of Finished Goods." Unpublished Ph.D. dissertation, the University of Minnesota.

Fair, Ray C. and John B. Taylor. (1980) "Solution and Maximum Likelihood Estimation of Dynamic Nonlinear Rational Expectations Models." National Bureau of Economic Research Technical Paper No. 5 (October).

Fischer, Stanley. (1979) "Anticipations and the Nonneutrality of Money." <u>Journal of Political Economy</u>, 87 (April), 225-252.

Futia, Carl A. (1981) "Rational Expectations in Stationary Linear Models." <u>Econometrica</u>, 49 (January), 171-192.

Gabel, Robert A. and Richard A. Roberts. (1973) <u>Signals and Linear Systems</u>. New York: John Wiley and Sons.

Gantmacher, F.R. (1959) <u>The Theory of Matrices</u> Vol. 1. New York: Chelsea Publishing Company.

Gourieroux, C.; Laffont, J.J.; and A. Monfort. (1982) "Rational Expectations in Dynamic Linear Models: Analysis of the Solutions." <u>Econometrica</u>, 50 (March), 409-426.

Granger, Clive W.J. (1969) "Investigating Causal Relations by Econometric Models and Cross-Spectral Methods." <u>Econometrica</u>, 37 (July), 424-438.

Hannan, E.J. (1970) <u>Multiple Time Series</u>. New York: John Wiley and Sons.

Hansen, Lars P. and Thomas J. Sargent. (1980a) "Formulating and Estimating Dynamic Linear Rational Expectations Models." <u>Journal of Economic Dynamics and Control</u>, 2 (February), 7-46.

_____. (1980b) "Methods for Estimating Continuous Time Rational Expectations Models from Discrete Data." Federal Reserve Bank of Minneapolis Staff Report No. 59 (April).

_____. (1980c) "Rational Expectations Models and the Aliasing Phenomenon." Federal Reserve Bank of Minneapolis Staff Report No. 60 (May).

_____. (1981a) "Linear Rational Expectations Models for Dynamically Interrelated Variables." <u>Rational Expectations and Econometric Practice</u>. Edited by R.E. Lucas Jr. and T.J. Sargent.

_____. (1981b) "Exact Linear Rational Expectations Models: Specification and Estimation." Federal Reserve Bank of Minneapolis Staff Report No. 71 (September).

Jacobs, Rodney L. (1975) "A Difficulty with Monetarist Models of Hyperinflation." <u>Economic Inquiry</u>, 13 (September), 337-360.

Kailath, Thomas. (1980) <u>Linear Systems</u>. Englewood Cliffs: Prentice-Hall.

Kennan, John. (1979) "The Estimation of Partial Adjustment Models with Rational Expectations." <u>Econometrica</u>, 47 (November), 1441-1455.

Kohn, R. (1979) "Asymptotic Estimation and Hypothesis Testing Results for Vector Linear Time Series Models." <u>Econometrica</u>, 47 (July), 1005-1030.

Kydland, Finn E., and Edward C. Prescott (1977) "Rules Rather Than Discretion: The Inconsistency of Optimal Plans." <u>Journal of Political Economy</u> 85 (June), 473-491.

Lucas, Robert E., Jr. (1972a) "Econometric Testing of the Natural Rate Hypothesis." <u>Econometrics of Price Determination Conference</u>. Edited by Otto Eckstein. Washington, D.C.: Board of Governors of the Federal Reserve System.

_____. (1972b) "Expectations and the Neutrality of Money." <u>Journal of Economic Theory</u>, 4 (April), 103-124.

_____. (1973) "Some International Evidence on Output-Inflation Tradeoffs." <u>American Economic Review</u>, 63 (June), 326-334.

_____. (1975) "An Equilibrium Model of the Business Cycle." <u>Journal of Political Economy</u>, 83 (December), 1113-1144.

_____. (1976) "Econometric Policy Evaluation: A Critique." <u>The Phillips Curve and Labor Markets</u>. Edited by K. Brunner and A.H. Meltzer, Carnegie-Rochester Conference on Public Policy. Amsterdam: North-Holland, 19-46.

_____ and Edward C. Prescott. (1971) "Investment Under Uncertainty." <u>Econometrica</u> 39 (September), 659-681.

_____ and Thomas J. Sargent. (1978) "After Keynesian Macroeconomics." <u>After the Phillips Curve: Persistence of High Unemployment</u>. Boston: Federal Reserve Bank of Boston.

_____. (1981) "Introduction." <u>Rational Expectations and Econometric Practice</u>. Edited by Robert E. Lucas, Jr. and Thomas J. Sargent. Minneapolis: University of Minnesota Press.

McCallum, Bennett T. (1980) "On Non-Uniqueness in Rational Expectations Models: An Attempt at Perspective." (May). (Mimeographed.)

Muth, John F. (1961) "Rational Expectations and the Theory of Price Movements." Econometrica, 29 (July), 315-335.

Nerlove, Marc; Grether, David M.; and Jose L. Carvalho. (1979) Analysis of Economic Time Series. New York: Academic Press.

Noble, Ben. (1969) Applied Linear Algebra. Englewood Cliffs: Prentice-Hall.

Pace, I.S. and S. Barnett. (1974) "Efficient Algorithms for Linear System Calculations. Part I--Smith Form and Common Divisor of Polynomial Matrices." International Journal of Systems Science, Vol. 5, No. 5, pp. 403-411.

Phadke, M.A. and G. Kedem. (1978) "Computation of the Exact Likelihood Function of Multivariate Moving Average Models." Biometrika, 65, 511-519.

Rozanov, Yu. A. (1967) Stationary Random Processes. San Francisco: Holden-Day.

Salemi, Michael K. (1979) "Adaptive Expectations, Rational Expectations, and Money Demand in Hyperinflation Germany." Journal of Monetary Economics, 5 (October), 593-604.

_____. (1981) "The Solution of Linear Rational Expectations Models." Manuscript, University of North Carolina at Chapel Hill.

_____ and T.J. Sargent. (1979) "The Demand for Money During Hyperinflation under Rational Expectations: II." International Economic Review, 20 (October), 741-758.

Saracoglu, Rusdu and Thomas J. Sargent. (1978) "Seasonality and Portfolio Balance Under Rational Expectations." Journal of Monetary Economics, 4 (August), 435-458.

Sargent, Thomas J. (1976) "Econometric Exogeneity and Alternative Estimators of Portfolio Balance Schedules for Hyperinflations." Journal of Monetary Economics, 2 (April), 511-521.

_____. (1977) "The Demand for Money During Hyperinflations Under Rational Expectations: I." International Economic Review, 18 (February), 59-82.

_____. (1978) "Estimation of Dynamic Labor Demand Schedules Under Rational Expectations." Journal of Political Economy, 86 (December), 1009-1044.

_____. (1979a) "A Note on Maximum Likelihood Estimation of the Rational Expectations Model of the Term Structure." Journal of Monetary Economics, 5 (January), 133-143.

_____. (1979b) Macroeconomic Theory. New York: Academic Press.

_____. (1980) "Lecture Notes on Linear Prediction and Control." Manuscript.

_____. (1981) "Interpreting Economic Time Series." Journal of Political Economy, 89 (April), 213-218.

___ and N. Wallace. (1973) "Rational Expectations and the Dynamics of Hyperinflation." *International Economic Review*, 14 (June), 328-350.

___. (1975) "Rational Expectations, the Optimal Monetary Instrument and the Optimal Money Supply Rule." *Journal of Political Economy*, 83 (April), 241-255.

___. (1976) "Rational Expectations and the Theory of Economic Policy." *Journal of Monetary Economics*, 2 (April), 169-184.

Shiller, Robert. (1972) "Rational Expectations and the Term Structure of Interest Rates." Unpublished Ph.D. thesis, Cambridge, Massachusetts: M.I.T.

___. (1978) "Rational Expectations and the Dynamic Structure of Macroeconomic Models: A Critical Review." *Journal of Monetary Economics*, 4 (January), 1-44.

Sims, Christopher A. (1972) "Money, Income, and Causality." *American Economic Review*, 62 (September), 540-552.

___. (1977) "Exogeneity and Causal Ordering in Macroeconomic Models." *New Methods in Business Cycle Research: Proceedings from a Conference*. Minneapolis: Federal Reserve Bank of Minneapolis.

Taylor, John B. (1977) "Conditions for Unique Solutions in Stochastic Macroeconomic Models with Rational Expectations." *Econometrica*, 45 (September), 1377-1385.

Wallace, Neil. (1980) "The Overlapping Generations Model of Fiat Money." *Models of Monetary Economies*. Edited by J. Kareken and N. Wallace. Minneapolis: Federal Reserve Bank of Minneapolis.

Wallis, Kenneth F. (1980) "Econometric Implications of the Rational Expectations Hypothesis." *Econometrica*, 48 (January), 49-73.

Whittle, Peter. (1963) *Prediction and Regulation by Linear Least Squares Methods*. Princeton: D. VanNostrand.

Wilson, Charles A. (1979) "Anticipated Shocks and Exchange Rate Dynamics." *Journal of Political Economy*, 87 (June), 639-647.

Author Index

Aoki, M. 4,30,35,38,44

Bain, I. 123
Barnett, S. 99
Black, F. 9,71
Blanchard, O. 4,11,38,55,57,90,95,101
Brock, W. 9,71
Brown, J. ix
Burmeister, E. 11

Cagan, P. 5
Canzoneri, M. 4,30,35,38,44
Carvalho, J. 81
Chow, G. 90,100
Christiano, L. 4
Chrystal, G. 108
Churchill, R. ix

de Branges, L. 66
Dornbusch, R. 12

Eckstein, Z. 12
Eichenbaum, M. 12

Fair, R. 90
Fischer, S. 55
Futia, C. 7,38,83,90

Gabel, R. 43
Gantmacher, F. 108
Gourieroux, C. 4,9,83
Granger, C. 25
Grether, D. 81

Hannan, E. 43
Hansen, L. 3,9,11,12,16,26,28,43,57,
93,99,112,113,114,116

Jacobs, R. 4

Kahn, C. 4,90,95,101
Kailath, T. 100,103,104
Kedem, G. 43
Kennan, J. 12
Kohn, R. 43
Kydland, F. 119

Laffont, J. 4,9,83
Lucas, R. 3,29,30,38,50,84,85,113,118

McCallum, B. 4,11,71,84
Monfort, A. 4,5,9,83
Muth, J. 30,37,38,44

Nerlove, M. 81
Noble, B. 21

Pace, I. 99
Phadke, M. 43
Prescott, E. 118,119

Roberts, R. 43
Rovnyak, J. 66
Rozanov, Y. 6,22

Salemi, M. 4,96
Saracoglu, R. 4,26,38,39,63
Sargent, T. 2,3,4,8,9,11,12,16,26,28
29,38,39,43,45,47,57,61,62,63,64,
93,99,112,113,114,116,117,119,120,
121,123
Shiller, R. 1,4,11,16,61,62,83

Sims, C. 10,25,28

Taylor, J. 9,71,74,78,85,87,90

Verhey, R. ix

Wallace, N. 4,29,87
Wallis, K. 38,61
Whittle, P. 22,40,78
Wilson, C. 12

Subject Index

Adaptive expectations 1
Analytic function 5,6,7,10,13,14,15,
   22,24,32,35,40,41,42,43,44,46,67,
   69,72,73,74,77,79,113,114,115,117
Annihilator operator 40
Autoregressive representation 8,13,28,
   42,50,64,68,72,82,96

Backward solution 2,38,55-61,70
Banach fixed point theorem 67
Bivariate representations 4,22-25,98
Blaschke's factors 11
Bounded sequence 2,3

Certainty-equivalence 116,118,119
Characteristic root 16,32,60,77,118,
   122
Coincident indicator 77
Competitive equilibrium 110,118,124
Consumer surplus 118
Contingency plan 120
Contour-integration 66
Contraction mapping 39,67
Convolution 43,50,65,96
Covariance generating function 43,
   63,81
Cross-equation restriction 8,22,28,
   70,88,116

Decision rule 114,118,119
Driving Process 1,3,5,11,12,22,72,
   79,85,86

Elementary operation 103
Euler equation 91
Exogeneity 1,26,28

Expectational difference equation
   1,4,5,11,16,22,29,31,33,38,39,
   44,46,54,61,62,66,69,70,72,110,
   112,113,121

Feedback 111,121
Fixed point 37,38,64,117
Forward solution 2,38,55-61,70
Fourier transform 43
Frequency domain 38,39,43,63,96,99
Functional equation 65,67
Fundamental process 3,6,10,14,22,
   24,25

Granger-causation 25,26,28,112

Hilbert space 6,66

Inexact model 26
Infinite series 2
Information set 5

L'Hopital's rule 83
Lag operator 6,29,40,41,47,61,62,69
Law of iterated expectations 61,
   100,112
Leading indicator 74-78
Leibniz's rule 112
Linear difference model 1
Linearly regular covariance stationary stochastic process
   (LRCSSP) 5,10,21,34,35,39,72,
   77,111
Lucas Critique 3

Matrix pencil 100

Maximum likelihood 43,70,88,99,
    123
Minimum state variable condition
    71,84-86,87
Minimum variance condition 71,74,
    78-84,85,87
Monetary equilibrium 87
Monic polynomial 103,104

Newton's formulas 108
Nonsingular process 25,29
Nonstationarity 4,10
Nonuniqueness 9,25,51,71-89
Normal rank 103,104

Operator methods 38,61-63,70
Optimal linear regulator 116
Ordinary difference equation 1,
    38,44,61

Parametric nonuniqueness 72-74,
    84,88
Parseval's relation 66
Partial fractions 47,68,69,93,
    99,107
Perfect foresight 9
Perturbed equation 4,24,26-29
Plussing operator 66
Polynomial matrix 90,93,96,100,
    103,106,107

Rational expectations equilibrium
    110,117
Realization 3,7,23,27,73,91,92
Reduced form 38
Regularity 22,23-25
Residue 7,14,15,18,25,33,41,69,
    115
Riccati difference equation 116,
    118,119
Riesz-Fischer theorem 41,44

Singularity 7,10,14,18,25,33,41,
    42,73,81
Smith normal form 99,104,105
Social planner 118,124
Solution Principle 3-4,5,6,8,9,12
Spectrum 43
Square-summable sequence 3,6,7,
    10,32,42,44,49,66,72
State-space 38,50
Stochastic difference equation
    1,11,37
Stochastic process 4,9,30,39,117

Time domain 44,50

Undetermined coefficients 36,38,
    39,43,44,50,55,64,67,84
Unimodular matrix 103-105

Wiener-Kolmogorov formula 3,6,7,
    17,23,40,73,92,116
Withholding equation 4,5,29-36,
    76,90,100,101
Wold representation 3,5,6,7,8,9,
    12,13,14,17,21,22,26,31,40,
    72,91,114

z-transform 3,7,10,13,14,17,18,
    23,27,31,32,33,34,36,43,49,
    58,63,64,65,66,73,75,77,78,
    85,86,92,94,101,113,115

Charles Whiteman earned his B.A. at the University of Kansas and his Ph.D. in economics at the University of Minnesota. He is an assistant professor of economics at the University of Iowa.